Art In Biosynthesis

The Synthetic Chemist's Challenge

Volume I

ACADEMIC PRESS RAPID MANUSCRIPT REPRODUCTION

Art in Biosynthesis

The Synthetic Chemist's Challenge

Volume I

Darshan Ranganathan Subramania Ranganathan
Department of Chemistry
Indian Institute of Technology Kanpur

Foreword by
D.H.R. Barton
Imperial College, London

ACADEMIC PRESS New York San Francisco London 1976
A Subsidiary of Harcourt Brace Jovanovich, Publishers

ACADEMIC PRESS, INC.
111 Fifth Avenue, New York, New York 10003

United Kingdom Edition published by
ACADEMIC PRESS, INC. (LONDON) LTD.
24/28 Oval Road, London NW1

Library of Congress Cataloging in Publication Data

Ranganathan, Darshan.
Art in biosynthesis.

Bibliography: p.
Includes indexes.
1. Biosynthesis. 2. Chemistry, Organic.
I. Ranganathan, Subramania, joint author. II. Title.
QH345.R36 574.1′929 76-4108
ISBN 0–12–580001–0

PRINTED IN THE UNITED STATES OF AMERICA

This Work Is Dedicated to the Cause of

A Greater Awareness of
Nature by Man,
with Fond Hopes for a Deeper Appreciation
of Her Presence in Our Midst and an Increased
Desire to Simulate Her Experience in Building
Molecular Frameworks

Contents

Preface

Art in Biosynthesis projects Nature as an organic chemist supreme, and presents representative examples whose biosynthetic pathways are known on the basis of experimentation. *Art in Biosynthesis*, like the earlier *Art in Organic Synthesis* pays tribute to the creativity associated with the construction of molecular frameworks. The style of presentation of this book is similar to that of *Art in Organic Synthesis* to enable a quick comparison of the specific synthetic strategies. Such a comparison cannot but profoundly influence us and lead the art of organic synthesis in a direction dictated by Nature, which would indeed be most desirable. Simple in her means but inexhaustively rich in imagination and application, Nature operates more deftly whereas man's conquests with his vast armory of reagents and an infinite choice of reaction conditions, although awe inspiring and vastly impressive, cannot create that gentle glow of utter fascination. There is no need for us to regret our inadequacies. Nature has been building carbon frameworks for the past five billion years, and has been producing the necessarily highly sophisticated assemblies incorporating the genetic apparatus for nearly three billion years. Comparatively, man's experience as a creator of molecular frameworks is nascent and herein lies our relative lack of versatility and astuteness in planning. For example, Nature produces nearly 1000 indole alkaloids with just two building blocks, namely, tryptophan and loganin! The chemical simulation of biological processes is an endeavor that has captured the imagination of the organic chemists in recent times. This book is intended to promote this highly desirable transition. The largely pictorial presentations are followed by, in addition to the author index and subject index, a glossary, and α-amino acid index, a reagent index and reaction-type index to enable ready retrieval of information. In a large number of cases, biological reaction mechanisms are interpreted on the basis of *in vitro* experience. These are provided by us to maintain continuity and to emphasize the fact that biological transformations strictly follow the laws of bond-forming and bond-breaking that have been deduced from experience with *in vitro* systems.

We would like to express our deep sense of gratitude to Sir Derek Barton for his encouragement and patronage, which has, time and again—buffeted as we are with the ever present perturbations—helped us to align ourselves firmly to academic goals. We express our sincere indebtedness to Professor M.V. George of our department for his ever available encouragement, patronage, indulgence, and understanding.

It is with great pleasure that we acknowledge our gratitude and indebtedness to Mr. Nihal Ahmad and Mr. D.S. Panesar who are entirely responsible for the production of this manuscript. These colleagues took as much pride in the book as ourselves and their exemplary dedication to the arduous task was to us a great encouragement. We thank Mr. V.A. Narayan for his meticulous compilation of the author index.

We are most grateful to Academic Press, for their interest in the work, which was mainly responsible for the rapid progress of the book.

Darshan Ranganathan
Subramania Ranganathan

Foreword

Organic chemists have always been interested in the biosynthesis of natural products. At first, the main purpose of biosynthetic theory was to aid in the deduction of structure. At a time when structural determination was very difficult, the aid of a biosynthetic hypothesis could be invaluable. Robinson's derivation of the structure of morphine is a pertinent example.

Much, however, has changed in organic chemistry in the last 30 years. First, thanks to physical techniques, especially X-ray crystallography, the determination of structure has become a trivial matter. Second, thanks to radioactive labeling, it is now possible to confirm, or to disprove, a biosynthetic hypothesis. Indeed, in favorable cases, every sequential step in a long biosynthetic sequence can be elucidated. Of course, not all cases are favorable, because of difficulties of transfer of metabolites through cell walls. However, much has been accomplished, especially with plants and with microorganisms.

Once a biosynthetic sequence has been determined, it can be compared with biosynthetic hypothesis. It is indeed gratifying when the two are in agreement and when hypothesis can be used to make yet further, verifiable, predictions.

An established biosynthetic sequence is a challenge to the synthetic organic chemist to imitate and to equal Nature. Imitation is, in many cases possible, but we are far from equaling Nature in the yield and stereospecificity of her processes.

Art in Biosynthesis is a work of appreciation and criticism of the beauty of biosynthetic processes and of the interaction between man (the organic chemist) and Nature. It contains numerous well-chosen examples and will be of value as a source book, as an inspiration to research, and even more perhaps, as an aesthetic work of appreciation. The two authors are, by ability and background, specially gifted to write such a book. I am sure that this work will be very well received by chemists and biochemists.

Sir Derek Barton, D.Sc., FRS,
Nobel Laureate
Hofmann Professor of Organic Chemistry
Imperial College of Science and Technology
London

AJMALICINE

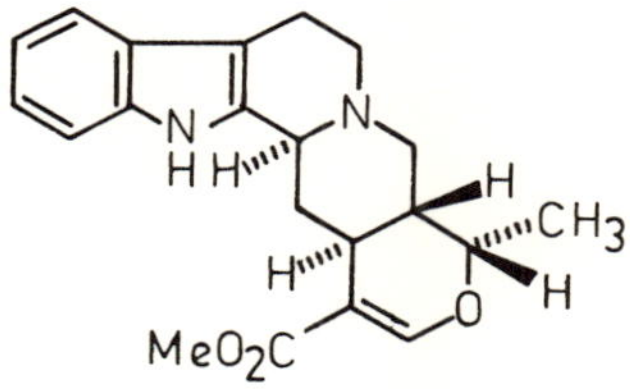

The synthesis of Ajmalicine from vincoside is an excellent illustration of Nature's subtle art of changing carbon frameworks by few, entirely rational transformations.[1]

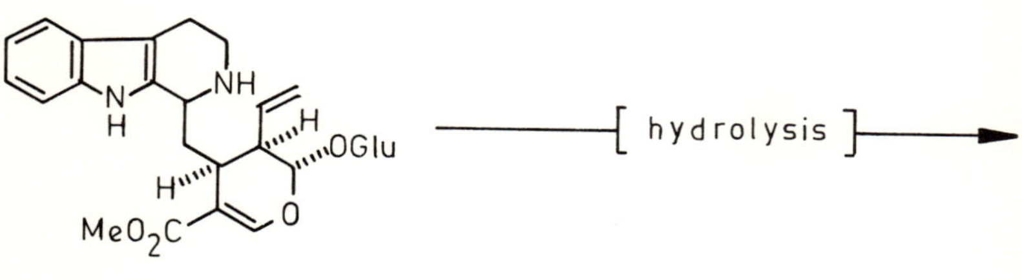

Vincoside(see p. 178)

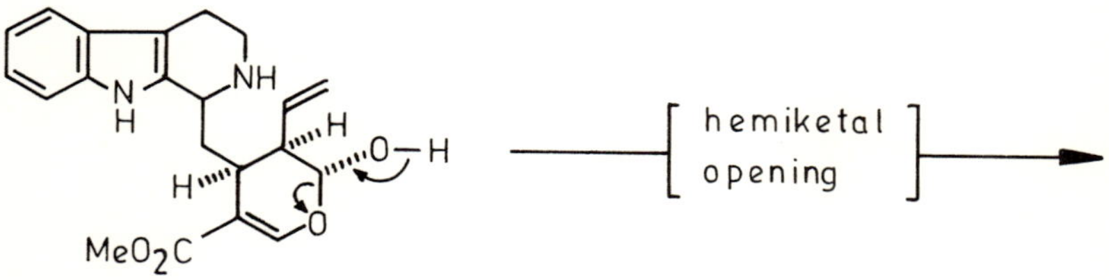

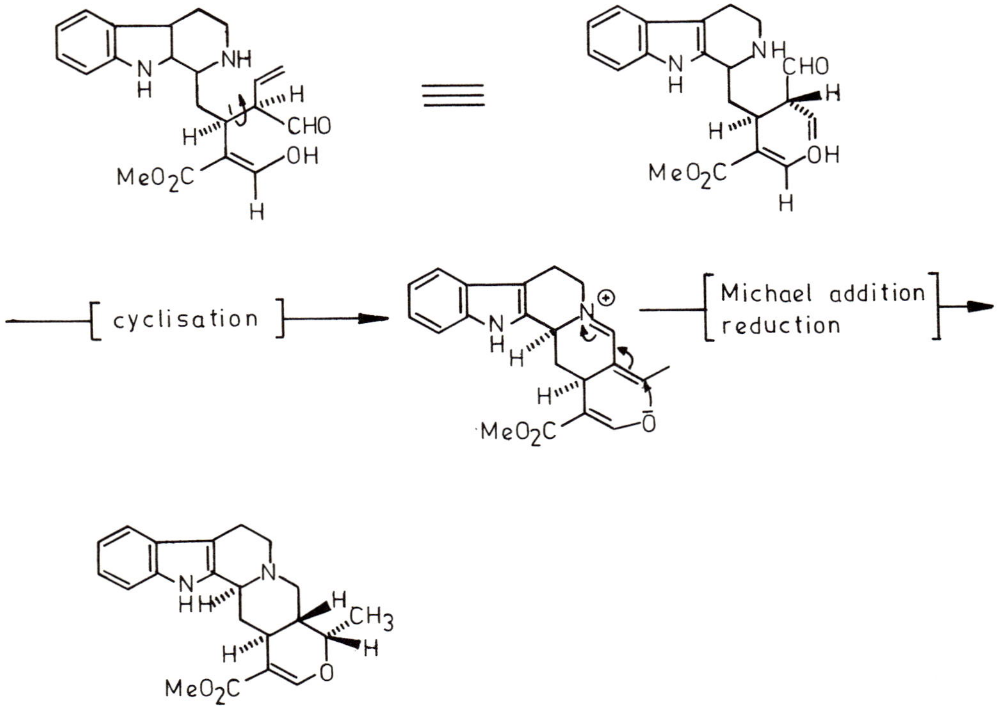

Ajmalicine

1. A. R. Battersby, Pure and Appl. Chem., 14, 117 (1967); A. R. Battersby, R. S. Kapil, J. A. Martin and L. Mo; Chem. Comm., 133 (1968); P. Loew and D. Arigoni, Chem.Comm., 137 (1968); A. R. Battersby, A. R. Burnett and P.G. Parsons, Chem. Comm., 1280 (1968); A. I. Scott, Acc. Chem. Res., 3 151 (1970); A. I. Scott, Bioorg. Chem., 3, 398 (1974).

AMMONIA NH_3

The biosynthesis of ammonia - nitrogen fixation - is, next to photosynthesis, the most important operation that is associated with life. Nearly all the nitrogen needed for growth is provided from biosynthesis with chemical fertilisers playing a minor role. The transformation of nitrogen to ammonia is a difficult process and in vitro is accomplished by hydrogenation at high temperatures and pressures. Yet, in Nature the same net reaction is achieved under ambient conditions! It was only in the last decade, after an intensive effort, the salient aspects of Nature's fascinating 'nitrogen-fixing' system became understood[1] leading to intense current efforts to simulate the biological nitrogen fixation.[2]

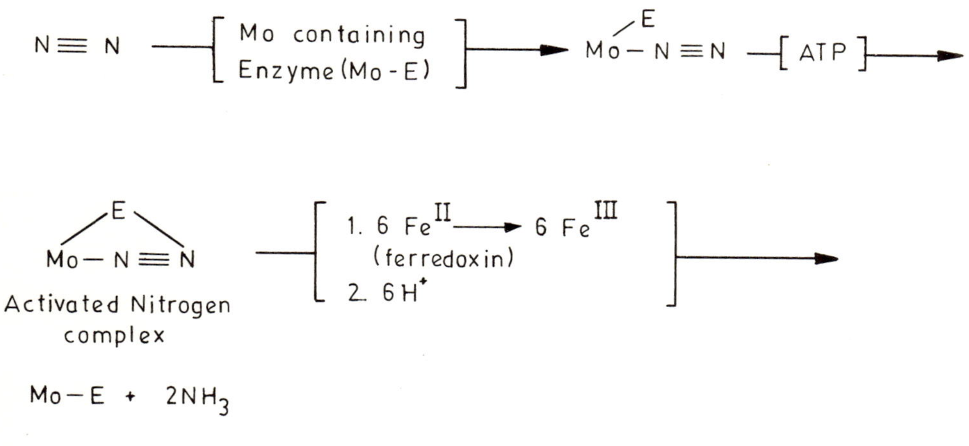

The ATP and the reduced ferredoxin that are involved, respectively, in the activation and reduction processes are created by Nature from pyruvic acid!

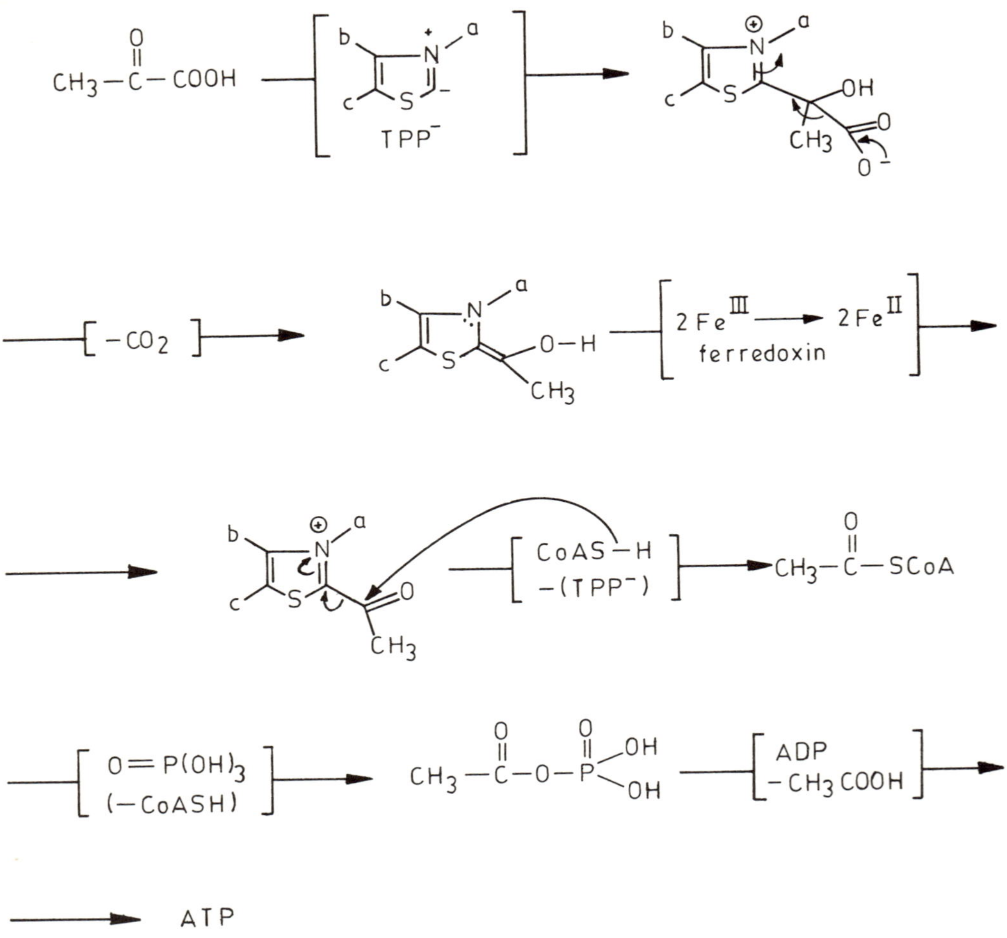

AMMONIA

1. D. Kleiner, Angew. Int., 14, 80 (1975).

2. J. G. Leigh and J. Chatt, Chem. Soc. Rev., 121 (1972); A. D. Allen, R.O. Harris, B. R. Loescher, J. R. Stevens and R. N. Whiteley, Chem. Rev., 73, 11 (1973).

β-AMYRIN

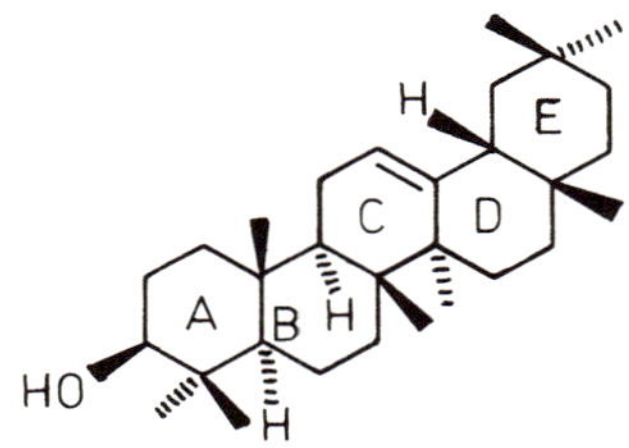

Nature generates several important carbon frameworks by folding squalene epoxide in different conformations. The Protolanosterol precursor to Cholesterol arises from a 'chair-boat-chair' arrangement whilst β -Amyrin is created by a 'chair-chair-chair' alignment![1-3]

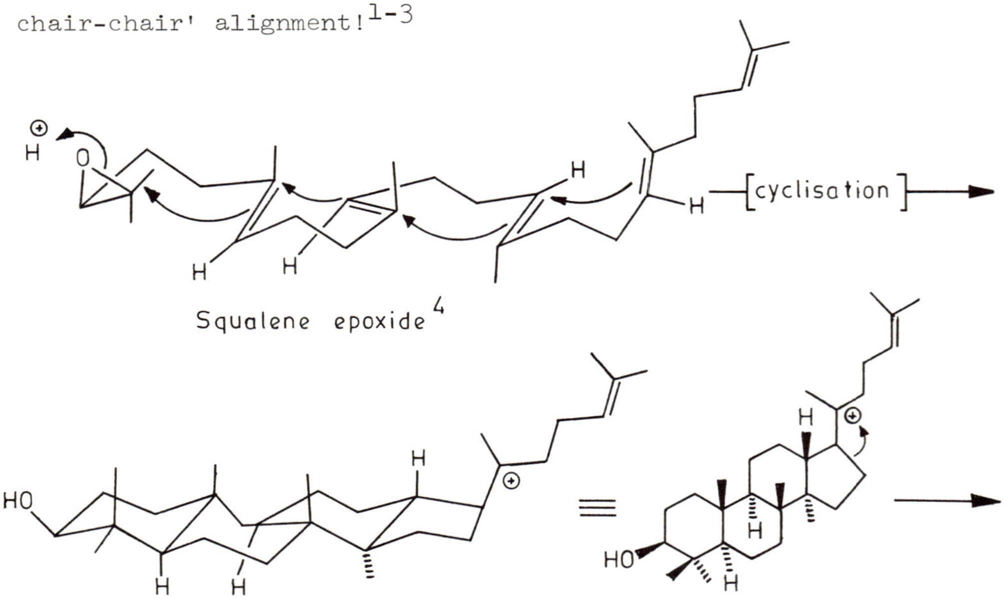

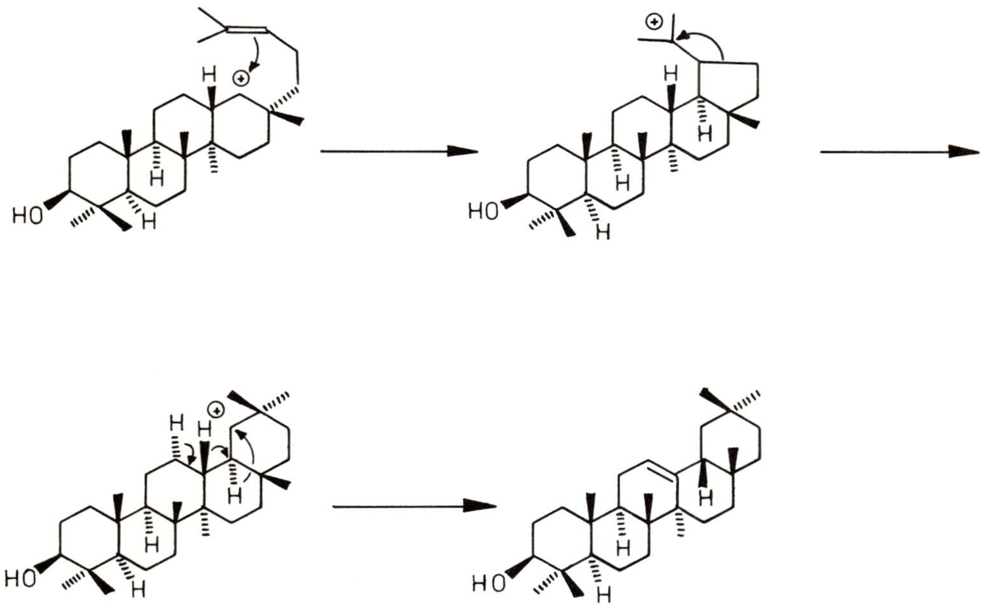

β- Amyrin

1. L. Ruzicka, Experientia, 9, 357 (1953); A. Eschenmoser, L. Ruzicka, O. Jeger and D. Arigoni, Helv. Chim. Acta, 38, 1890 (1955).

2. Experiments with labelled precursors have confirmed Ruzicka's hypothesis (H. H. Rees. G. Britton and T. W. Goodwin, Biochem. J., 103, 52 (1967); H. H. Rees, G. Britton and T. W. Goodwin, Biochem. J., 106, 659 (1968); D. H. R. Barton, G. Mellows, D. A. Widdowson and J. J. Wright, J. Chem. Soc., (C), 1142 (1971)).

3. ^{13}C-Labelling studies with [4-^{13}C] mevalonic acid have verified this hypothesis experimentally (S. Seo, Y. Tomita and K. Tori, Chem. Comm., 270 (1975)).

4. Recently, (S)-squalene 2,3 oxide was found to be the exclusive precursor of β -amyrin in plant systems (D. H. R. Barton, T. R. Jarman, K. G. Watson and D. A. Widdowson, Chem. Comm., 861 (1974)).

ANTHRAMYCIN

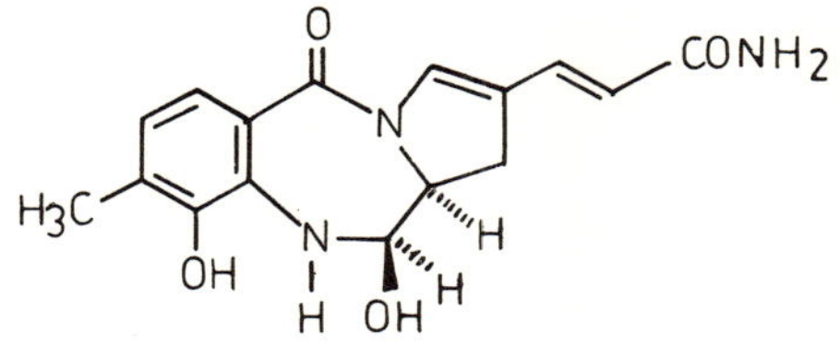

The construction of the Anthramycin framework by Nature, highlights, above all, the gainful utilisation of the metabolic products of tryptophan and tyrosine <u>via</u> a fascinating sequence and involving the unique transformation of a methyl group donated by methionine into an amide function![1]

Anthramycin - an antitumor agent possessing antibiotic, amebicidal and chemosterilant properties - is known to act by virtue of its ability to bind to DNA and therefore interfere with the function of DNA, thus causing inhibition of DNA and RNA synthesis in both bacterial and mammalian cells.

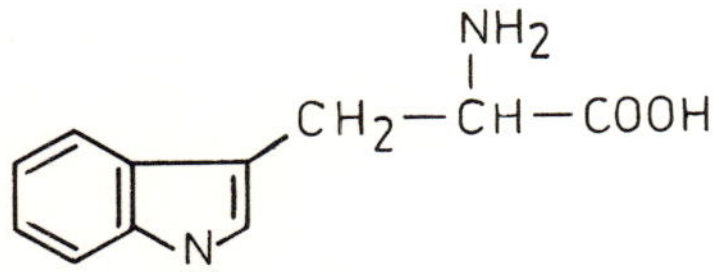

tryptophan (see p.173)

O_2 (oxygen)

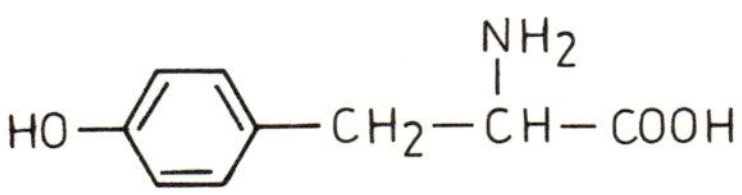

tyrosine[4] (see p.121)

HO^+

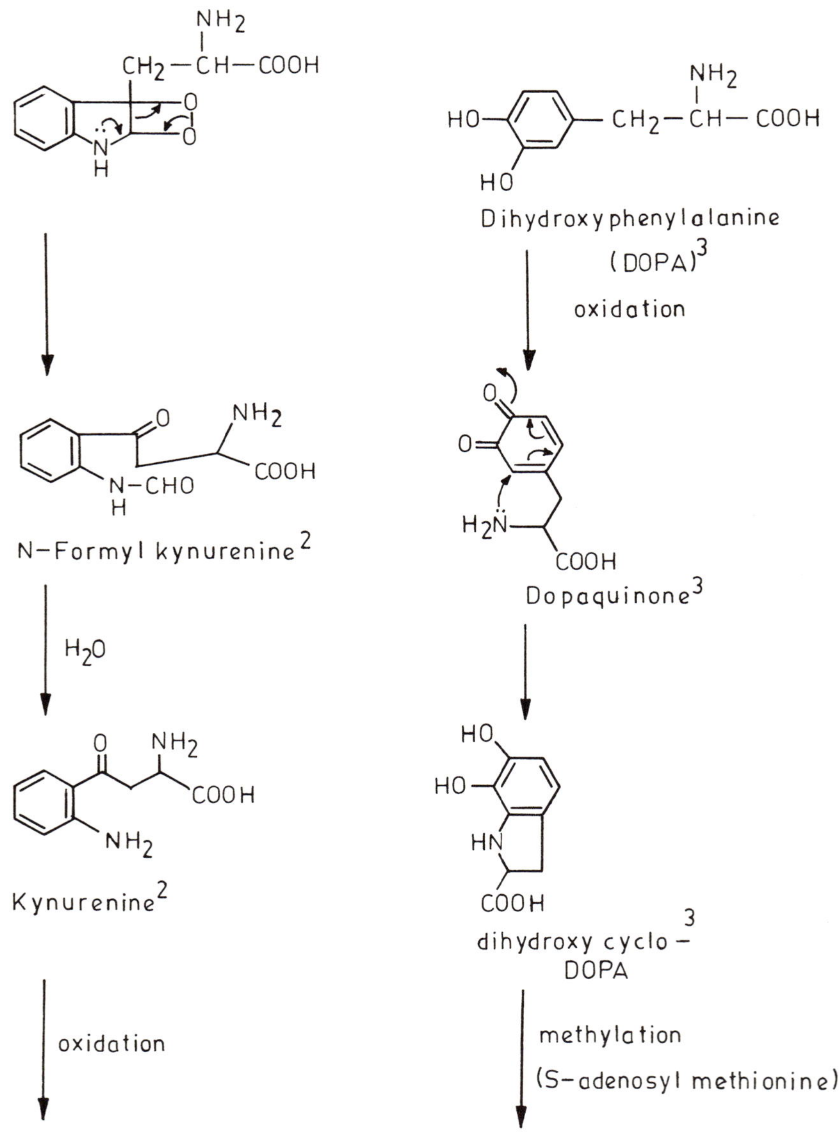
N-Formyl kynurenine[2]
H_2O
Kynurenine[2]
oxidation
Dihydroxyphenylalanine
(DOPA)[3]
oxidation
Dopaquinone[3]
dihydroxy cyclo-DOPA[3]
methylation
(S-adenosyl methionine)

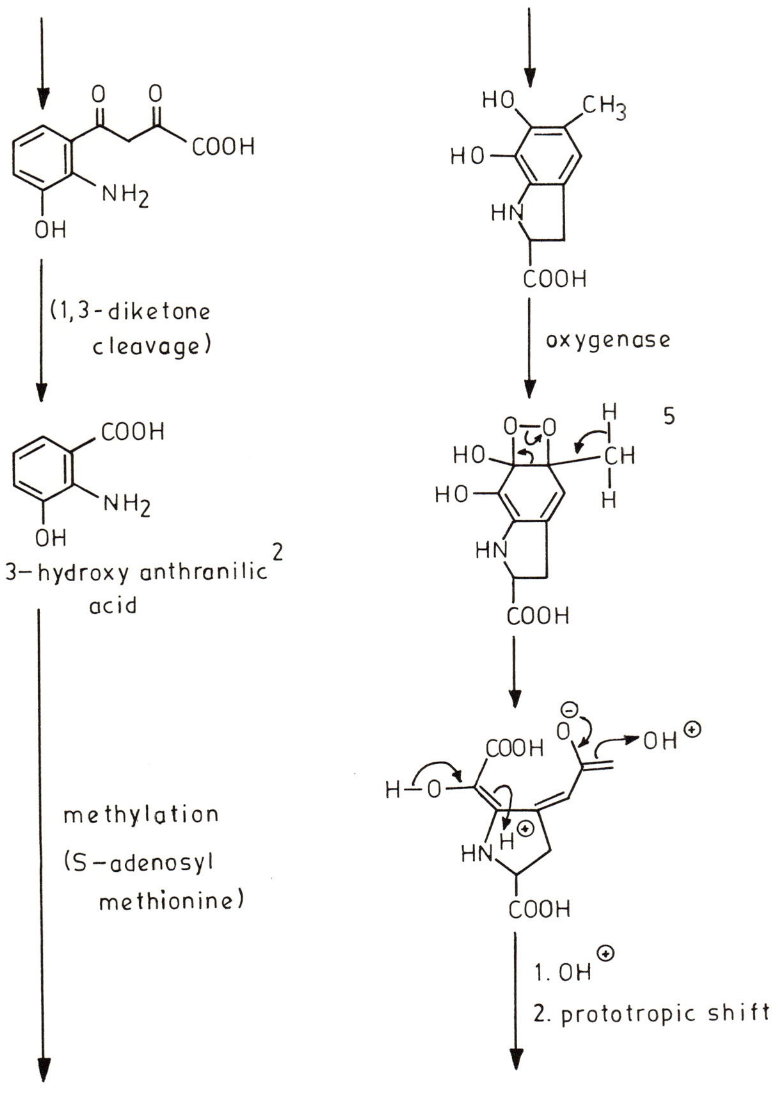
COOH
NH2
OH
(1,3-diketone cleavage)
3-hydroxy anthranilic acid
2
methylation
(S-adenosyl methionine)
HO
CH3
HN
oxygenase
5
COOH
1. OH
2. prototropic shift

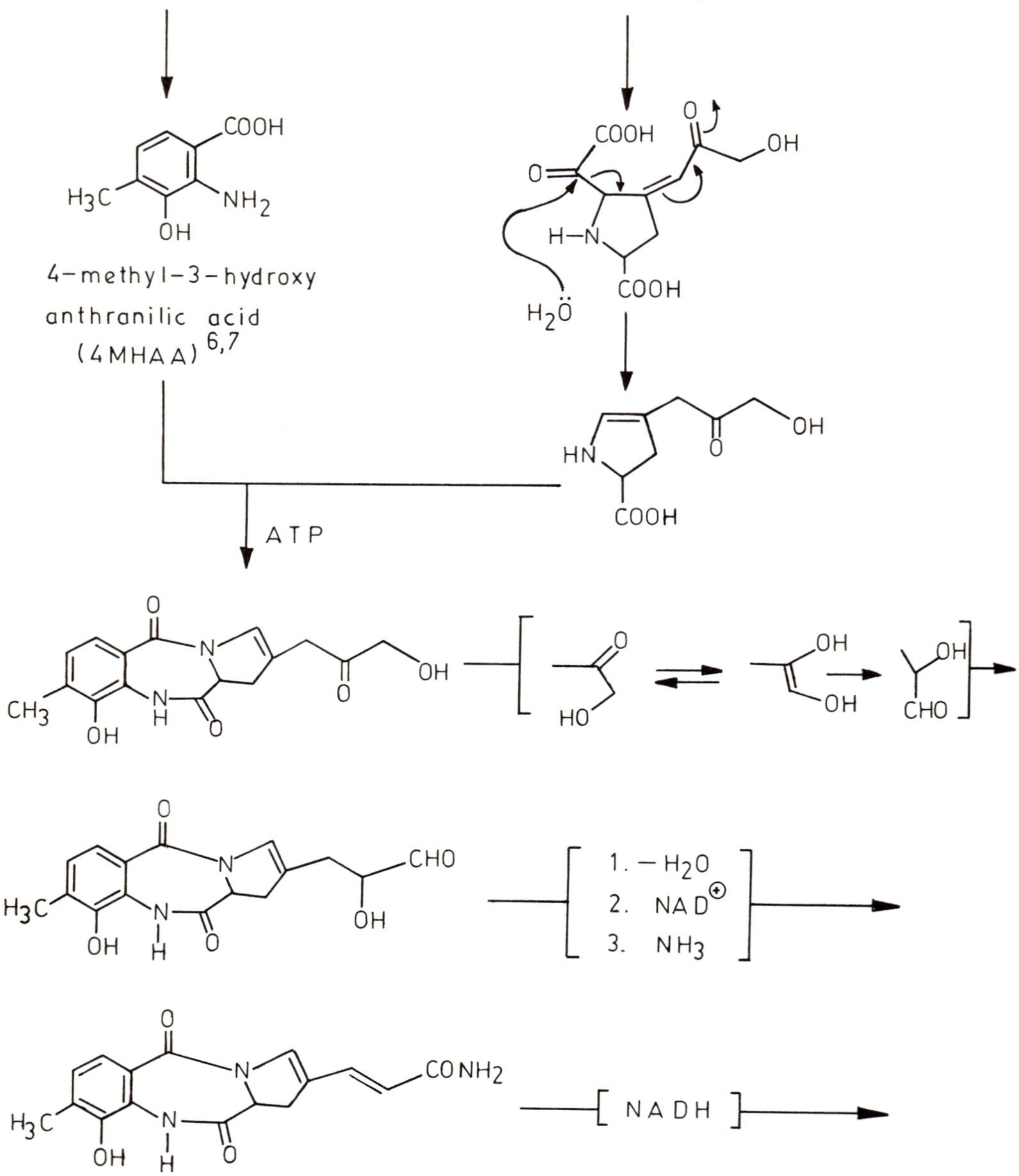
COOH
H_3C
NH_2
OH
4-methyl-3-hydroxy anthranilic acid (4MHAA) 6,7
COOH
O
OH
H-N
$H_2\ddot{O}$
COOH
HN
OH
O
COOH
ATP
O
N
OH
CH_3
N
H
OH
HO
OH
OH
OH
CHO
CHO
H_3C
1. $-H_2O$
2. $NAD^{\oplus}$
3. NH_3
$CONH_2$
NADH

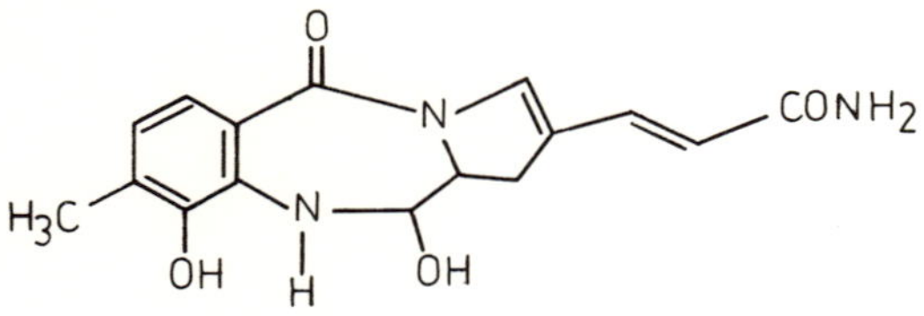

Anthramycin[8]

---

1. L. H. Hurley and M. Zmijewski, Chem. Comm., 337 (1974); L. H. Hurley, M. Zmijewski and C. J. Chang, J. Amer. Chem. Soc., 97, 4372 (1975).

2. Intermediate in the metabolism of tryptophan.

3. Intermediate in the metabolism of tyrosine.

4. Incorporation studies with L-[1-^{13}C], L-[U-^{14}C]-, L-[1-^{14}C]-tyrosine in anthramycin in *S. refuineus* showed that tyrosine contributes seven of its nine carbons to the C_3-proline moiety which arises in a manner analogous to that of propyl proline moiety of lincomycin A (D. F. Witz, E. Z. Hessler and T. L. Miller, Biochemistry, 10, 1128 (1971)).

5. Labelling pattern of anthramycin from L-[3,5-^{2}H]tyrosine confirmed the meta cleavage of the diol (Ref. 1).

6. Degradation of the labelled anthramycin showed that tryptophan is incorporated as 4-methyl-3-hydroxy-anthranilic acid, the methyl group originating from melthionine (Ref. 1).

7. The biosynthesis of 4-methyl-3-hydroxy anthranilic acid is analogous to that in actinomycin biosynthesis (L. Salzmann, H. Weissback and E. Katz, Arch. Biochem. Biophys., 30, 536 (1969); R. B. Herbert, Tetrahedron Letter, 4525 (1974)).

8. Incorporation of [Me-^{13}C] methionine showed that the [Me-^{13}C] methionine enriched the aromatic methyl group and the terminal amide carbonyl of the antibiotic. This probably is the first example of an amide carbonyl biogenetically derived from the S-methyl group of methionine.

BERBERINE

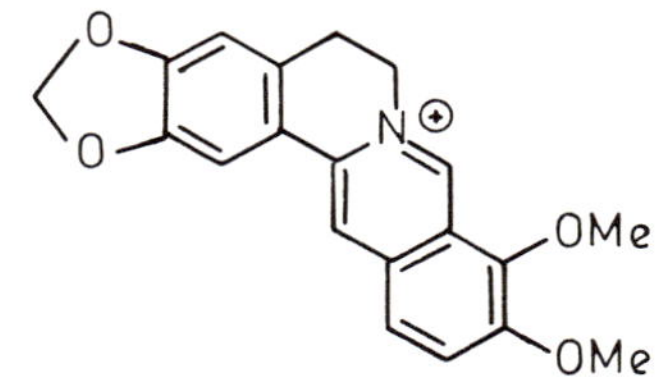

The biosynthetic transformation of reticuline to berberine is interesting in the sense that it demonstrates how a methyl group could be rationally incorporated into an aromatic system by Nature.[1] Another rather unique feature is the involvement of an N-oxide to achieve this[2] purpose!

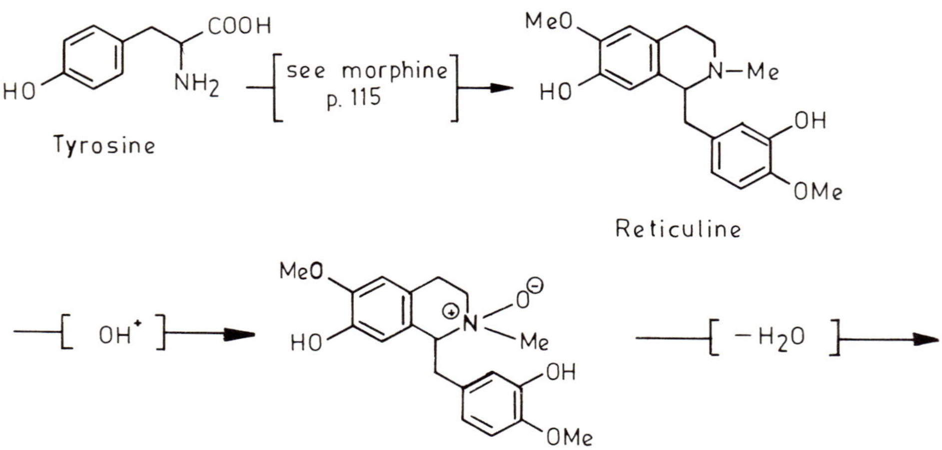

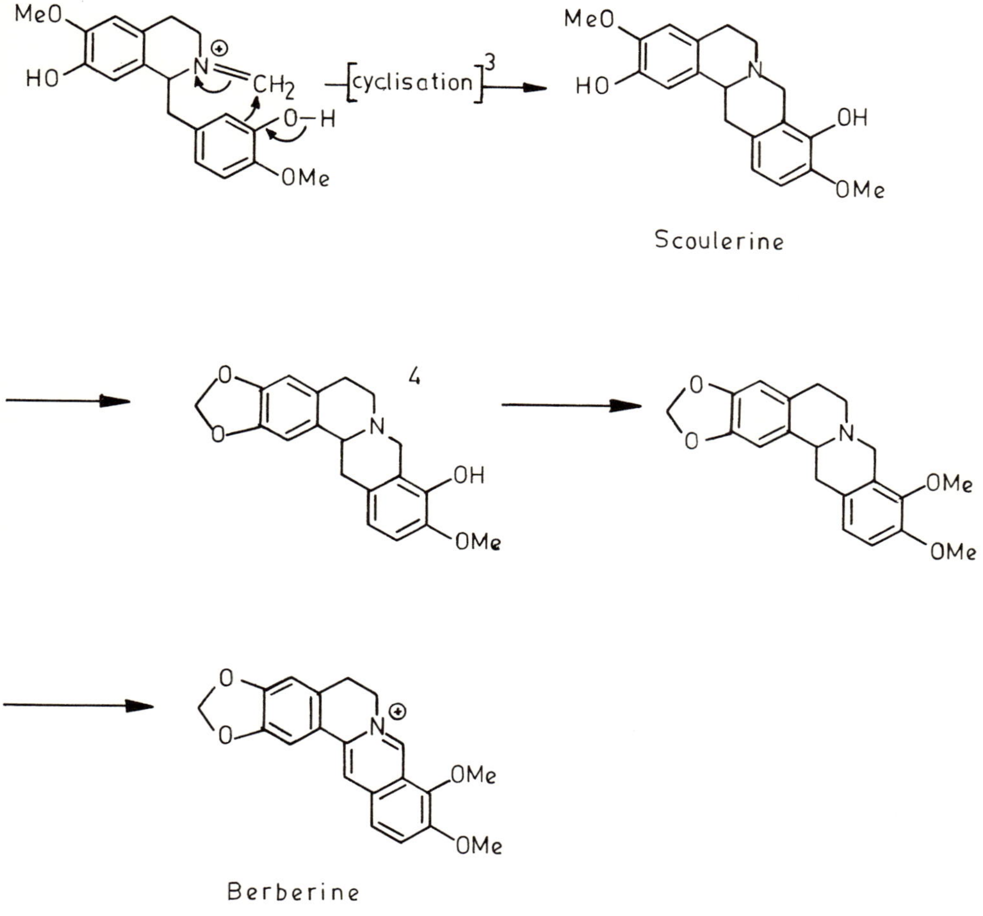

1. D. H. R. Barton, R. H. Hesse and G. W. Kirby, Proc. Chem. Soc., 267 (1963); D. H. R. Barton, Proc. Chem. Soc., 293 (1963); A. R. Battersby, Proc. Chem. Soc., 189 (1963); A. R. Battersby, R. J. Francis, M. Hirst and J. Staunton, Proc. Chem. Soc., 268 (1963); D. H. R. Barton, R. H. Hesse

and G. W. Kirby, J. Chem. Soc., 6379 (1965).

2. A. R. Battersby, Proc. Chem. Soc., 189 (1963).

3. The _in vitro_ analogy for this transformation is provided by treatment of the tertiary amine N-oxide with sulfurdioxide to form protoberberine (P. A. Bather, J. R. L. Smith and R. O. C. Norman, J. Chem. Soc., _C_, 3060 (1971)).

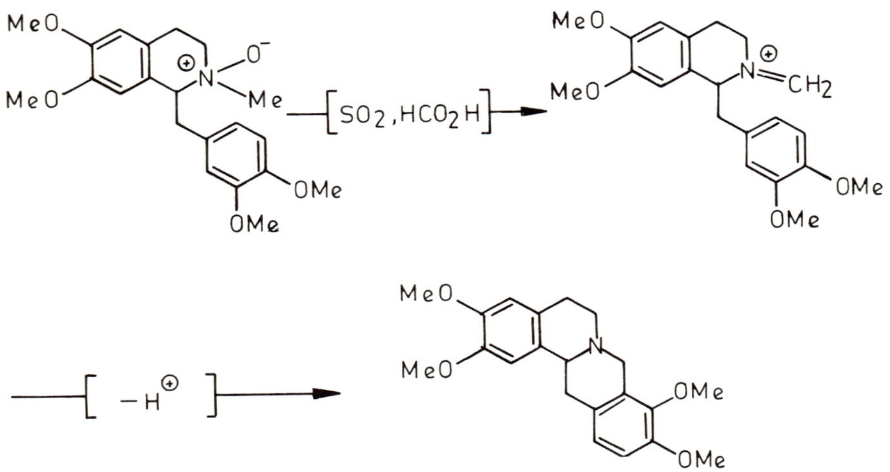

4. The methylenedioxy group is formed by oxidative cyclisation of an O-methoxyphenol function (D. H. R. Barton, G. W. Kirby, J. B. Taylor and G. M. Thomas, J. Chem. Soc., 4545 (1963)).

CAMPTOTHECIN

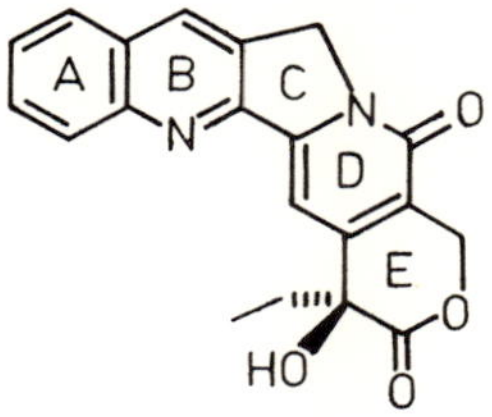

The anti-leukaemia and anti-tumor substance camptothecin[1] possesses an unusual quinoline framework which has now been shown to arise by an oxidative cyclization and fission of the indole unit in vincoside.[2] This key feature has been incorporated in an imaginative synthesis of camptothecin.[3]

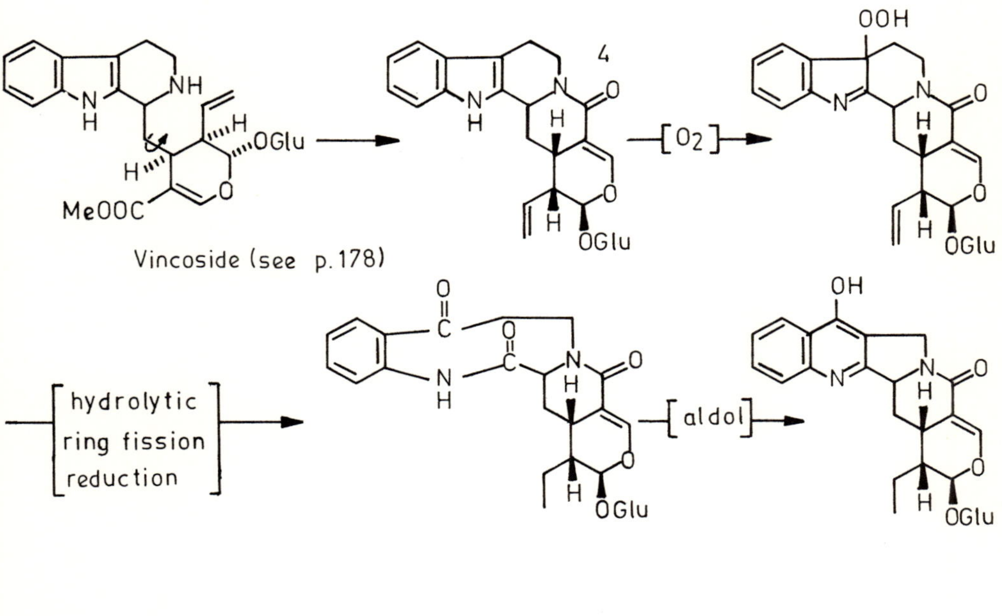

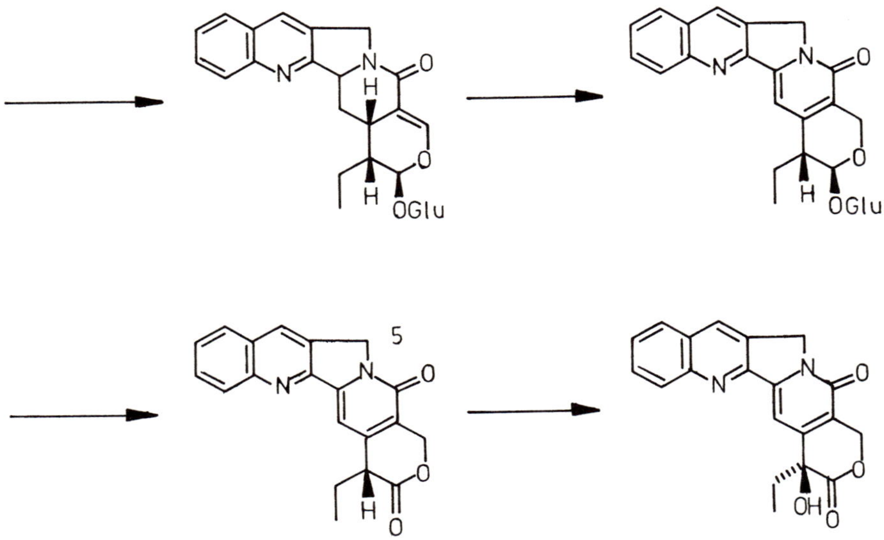

Camptothecin

1. A. G. Schultz, Chem. Rev., 73, 385 (1973).

2. C. R. Hutchinson, A. H. Heckenderf, P. E. Daddona, E. Hagaman and E. Wenkert, J. Amer. Chem. Soc., 96, 5609 (1974); E. Winterfeldt, Ann., 745, 23 (1971)

3. E. Winterfeldt, Angew. Chem., 84, 265 (1972).

4. This indole → quinoline change was postulated early. (E. Wenkert, K. G. Dave, R. G. Lewis and P. W. Sprague, J. Amer. Chem. Soc., 89, 6741 (1967); E. Winterfeldt, Ann., 745, 23 (1971)).

5. This hydroxylation could be chemically achieved (ref. 3).

CATHARANTHINE

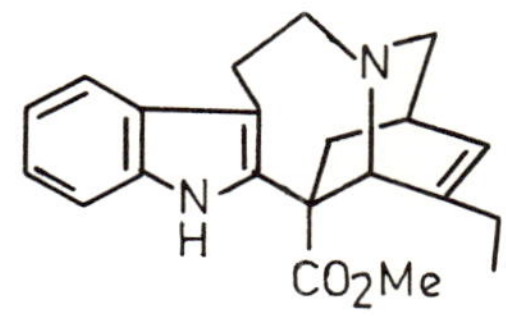

Catharanthine is created in Nature[1] from vincoside incorporating the fascinating manouvre of transfering the Loganin-β carbomethoxyhydroxymethylene unit to an indole α position followed by a fragmentation, [4+2] addition sequence.

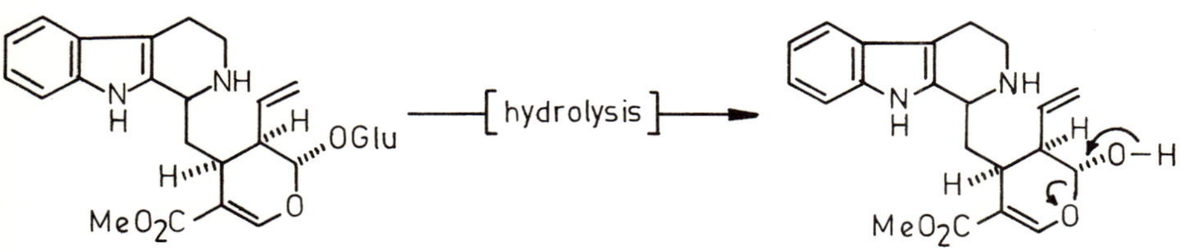

Vincoside (see p. 178)

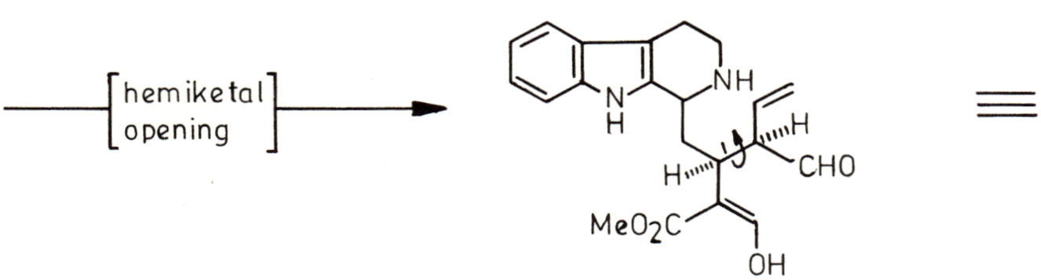

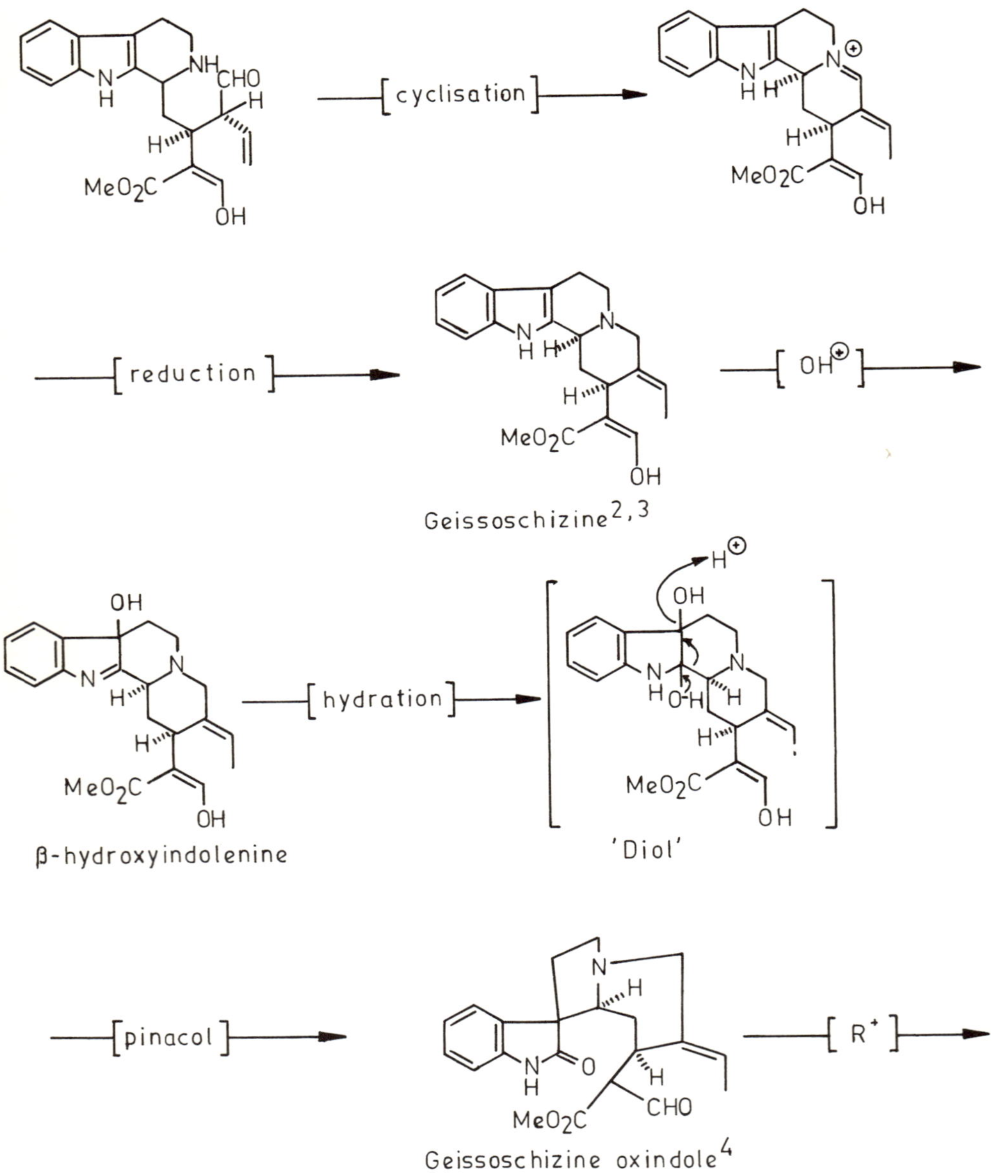
cyclisation
reduction
Geissoschizine[2,3]
OH⊕
β-hydroxyindolenine
hydration
'Diol'
pinacol
Geissoschizine oxindole[4]
R⁺

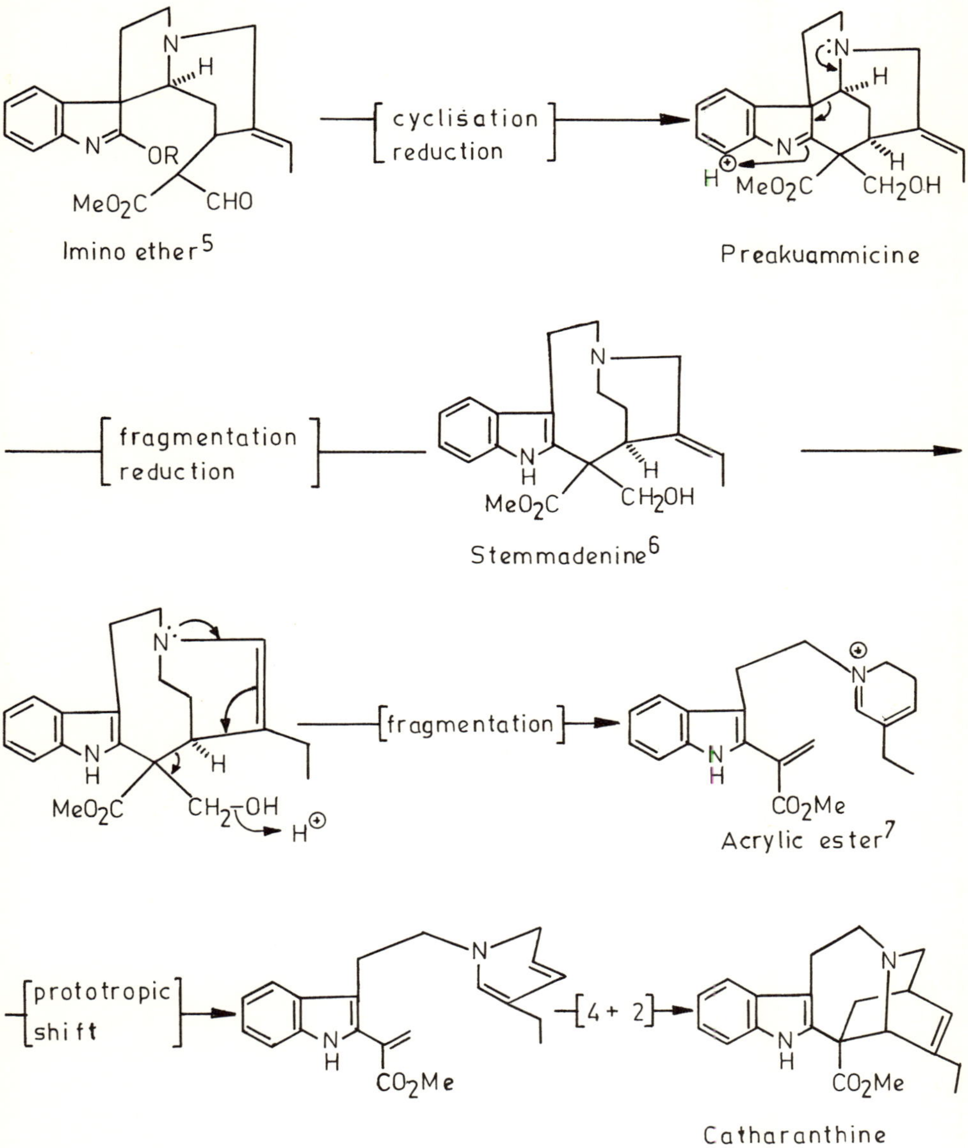
Imino ether[5]
cyclisation
reduction
Preakuammicine
fragmentation
reduction
Stemmadenine[6]
fragmentation
Acrylic ester[7]
prototropic
shift
[4 + 2]
Catharanthine

1. A. R. Battersby, Pure and Appl. Chem., 14, 117 (1967); A. I. Scott, Acc. Chem. Res., 3, 151 (1970); A. I. Scott, 'Alkaloids', M. T. P. International Review of Science, Organic Chemistry Series One, Vol. 9, eds. D.H. Hey and K. F. Weisner, Butterworths, University Park Press, 1973, p. 105; A. I. Scott, Bioorg. Chem., 3, 398 (1974).

2. In *Vinca rosea*, [O-methyl-^{3}H], [Ar-^{3}H] Geissoschizine was incorporated in catharanthine. (A. R. Battersby and E. S. Hall, Chem. Comm., 793 (1969).)

3. The sequence from Geissoschizine to Stemmadenine involving β-oxidation and oxindole chemistry, first proposed by Scott, has some preliminary feeding and isolation evidence (A. I. Scott, Acc. Chem. Res., 3, 151 (1970)).

4. Geissoschizine oxindole has been isolated from *Vinca rosea* and a labelled specimen incorporated into akuammicine.

5. The *in vitro* analogy for the proposed sequence involving oxindole and iminoether has been provided by the use of t-butyl-hypochlorite (A. I. Scott, Bioorg. Chem., 3, 398 (1974)).

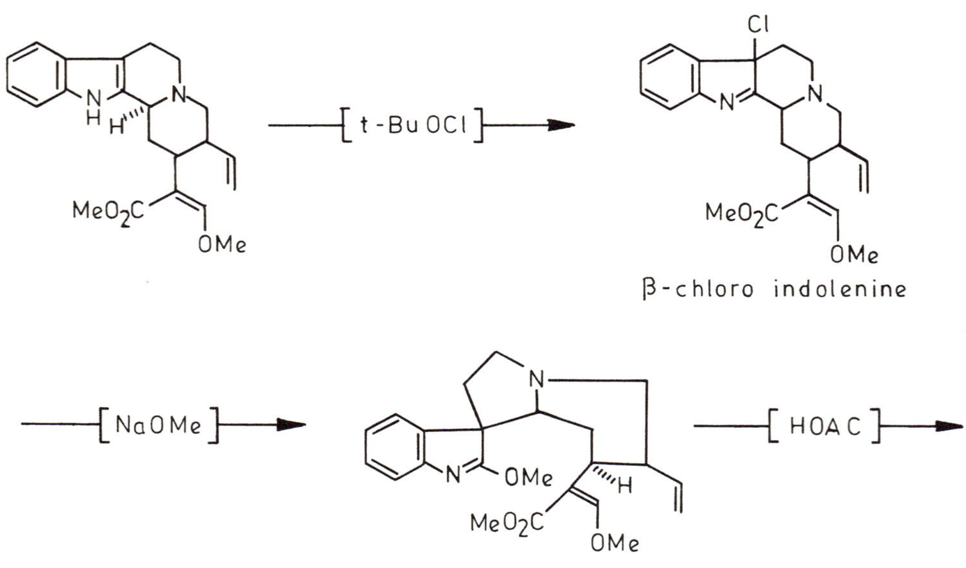

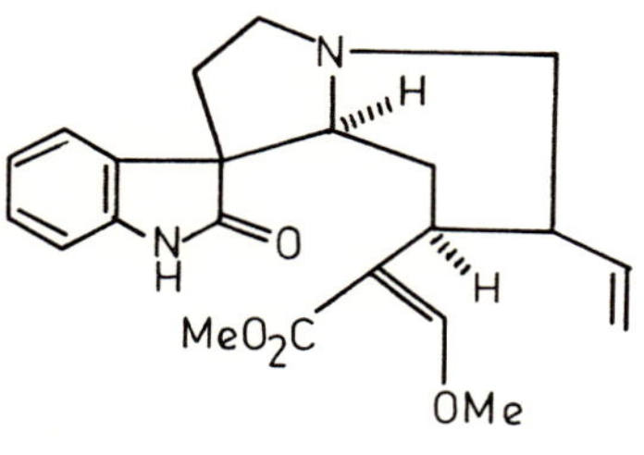

Corynoxeine

6. Stemmadenine has been converted into catharanthine in acetic acid at reflux (A. A. Qureshi and A. I. Scott, Chem. Comm., 945 (1968)).

7. The in vitro evidence for the formation of this acrylic ester has been obtained (A. I. Scott and P. C. Cherry, J. Am. Chem. Soc., 91, 5672 (1969)).

CECROPIA JUVENILE HORMONE

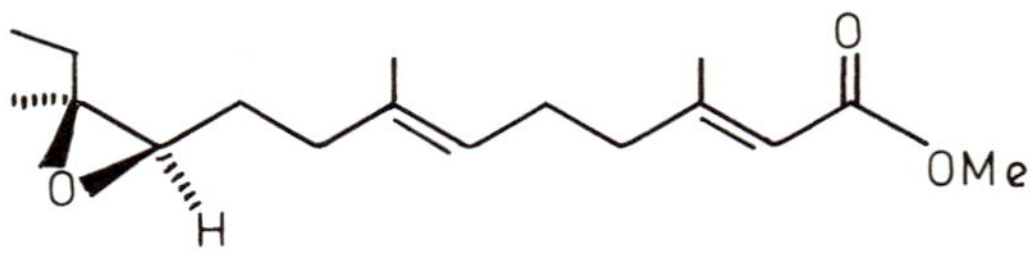

The biosynthesis[1,2] of cecropia Juvenile Hormone (JH II) illustrates an appealing facet of the art in biosynthesis relating to the gainful utilisation of a rather rare metabolic product, namely, propionyl-CoA, arising from degradation of odd carbon fatty acids.

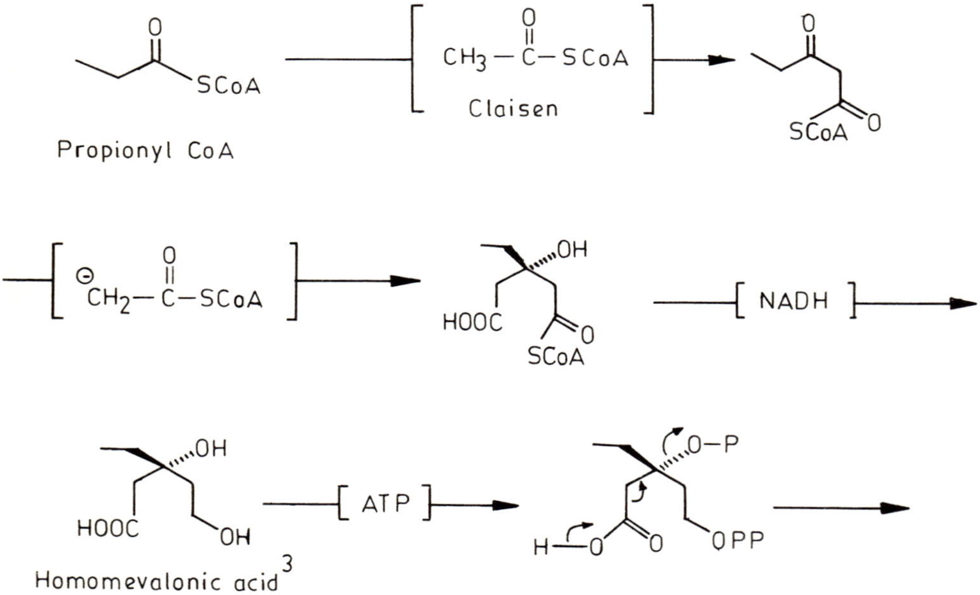

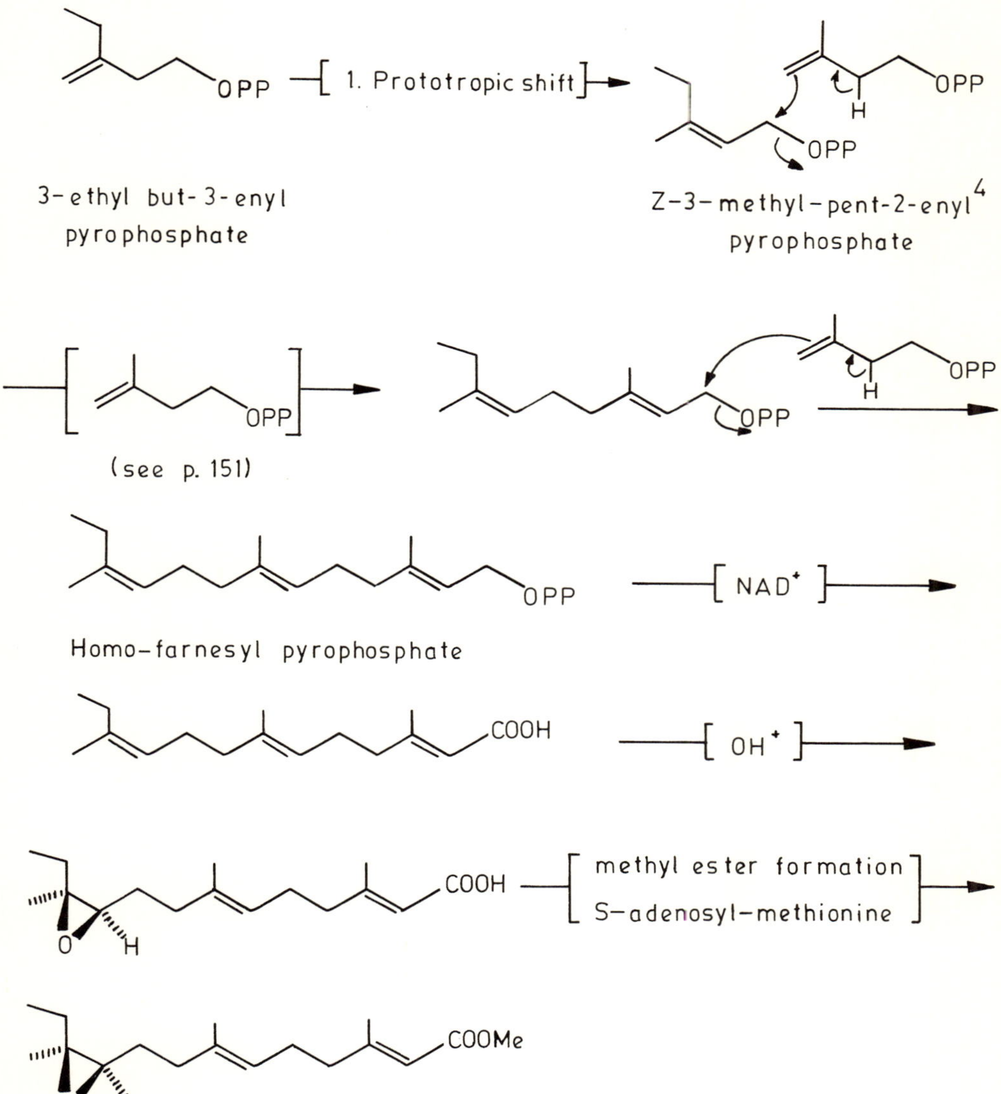
OPP
1. Prototropic shift
H
OPP
OPP
3-ethyl but-3-enyl pyrophosphate
Z-3-methyl-pent-2-enyl[4] pyrophosphate
OPP
(see p. 151)
OPP
H
OPP
OPP
NAD+
Homo-farnesyl pyrophosphate
COOH
OH+
COOH
O
H
methyl ester formation
S-adenosyl-methionine
COOMe
O
H
JH II

1. D. A. Schooley, K. J. Judy, B. J. Bergot, M. S. Hall and J. B. Siddall; Proc. Nat. Acad. Sci. USA, 70, 2921 (1973).

2. Labelled acetate, propionate and mevalonate were efficiently incorporated into JH II.

3. Radioactivity from [5-^{3}H] homomevalonate was incorporated into JH II (R. C. Jennings, K. J. Judy and D. A. Schooley, Chem. Comm., 21 (1975).

4. 3-Methyl-pent-2-enyl pyrophosphate was incorporated into homofarnesyl pyrophosphate in the presence of prenyl transferase (G. Popjak, J. L. Rabinowitz and J. M. Baron, Biochem. J., 113, 861 (1969); K. Ogura, T. Nishino, T. Koyama, and S. Seto, J. Amer. Chem. Soc., 92, 6036 (1970)).

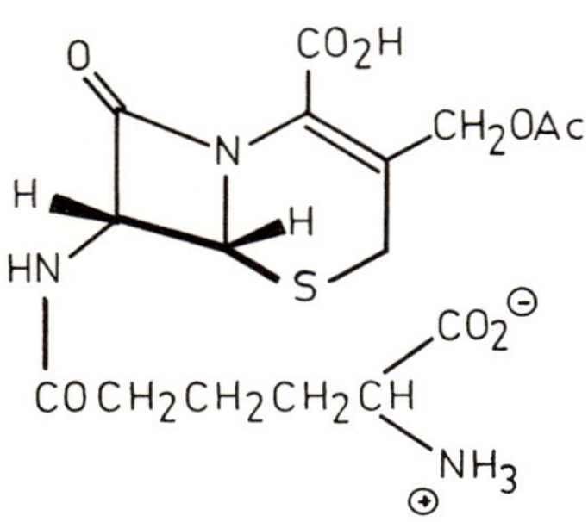

Cephalosporin-C

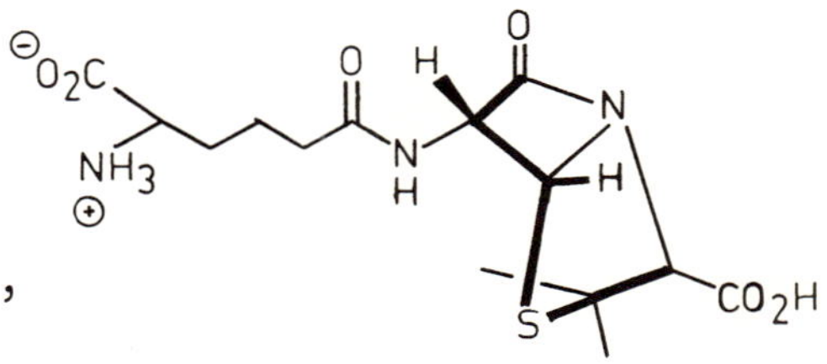

Penicillin-N

The synthesis and biosynthesis of cephalosporins and penicillins reflect the virtuosity of man and Nature as creators of delicate molecular frameworks.[1]

The structural elucidation and the synthesis of these antibiotics were attended with many frustrations and indeed at one time the idea that the penicillin framework cannot be properly defined was seriously entertained![2].

The biosynthesis of penicillins and cephalosporins is achieved by an ingenious plan from the three amino acids, L-α-aminoadipic acid, L-cysteine and L-valine.[3] Man's endeavour to unravel the finer details of Nature's synthetic plan - unlike most other natural products - have been thus far infructuous. The dipeptides and the tripeptide resulting from the three amino acids have been programmed, <u>via</u> different patterns, leading to penicillins and then cephalosporins. Many of such theories have good <u>in vitro</u> experimental support. The simplest of these is presented in the main text and the others are incorporated in the footnotes.[11,12] Penicillin S-oxide appears to be the most suited precursor for rearrangement leading to cephalosporins.

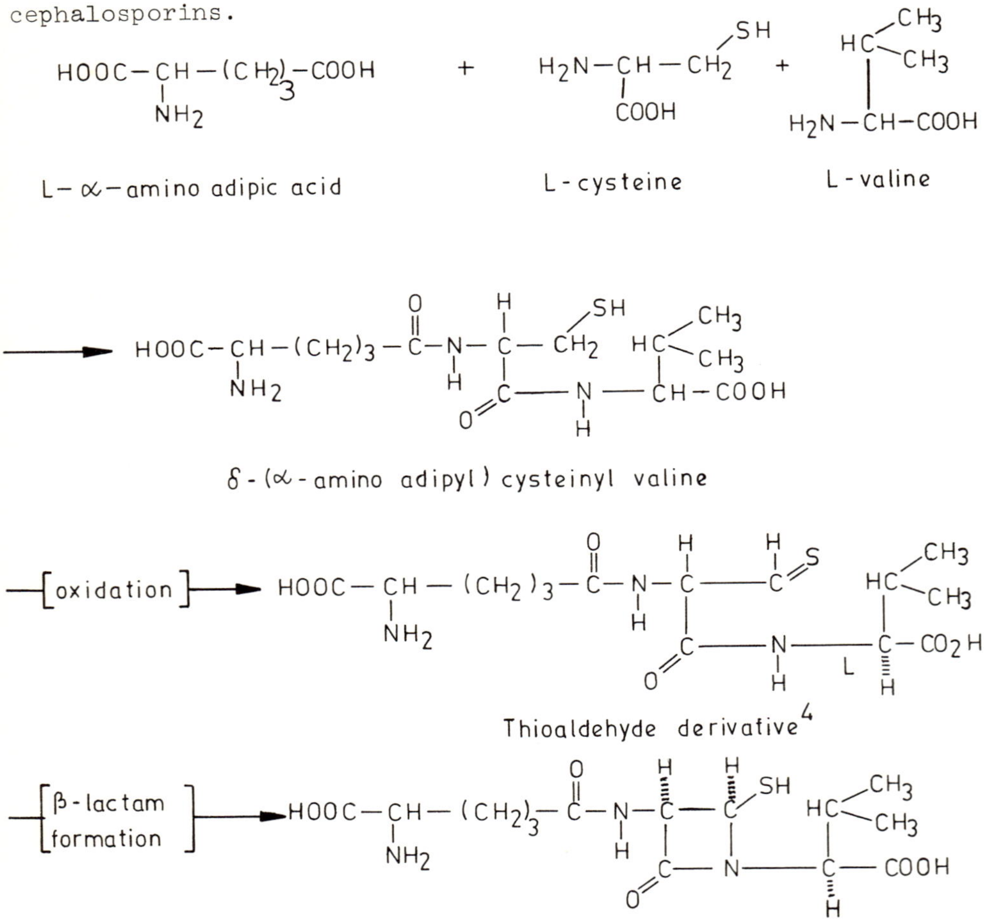

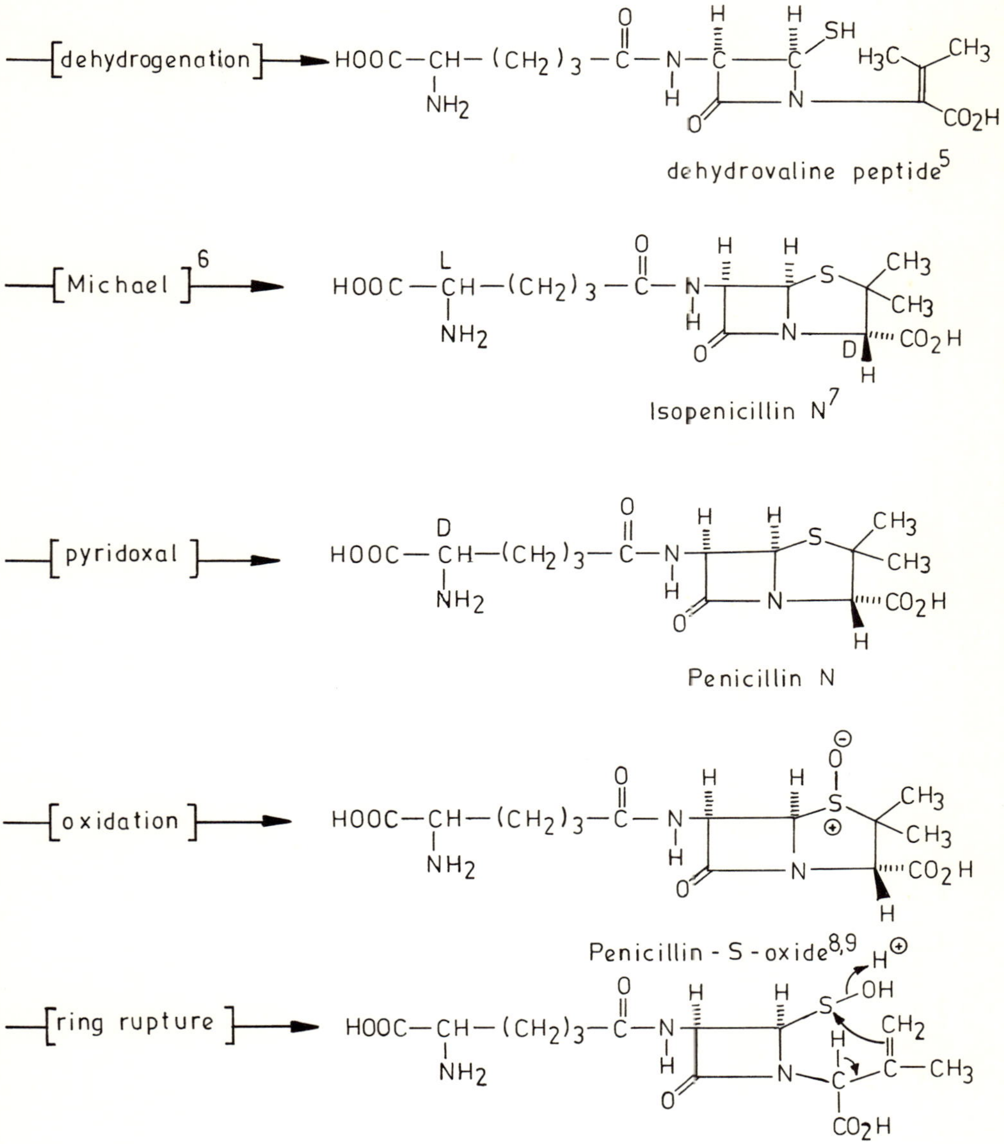
dehydrogenation
dehydrovaline peptide[5]
Michael[6]
Isopenicillin N[7]
pyridoxal
Penicillin N
oxidation
Penicillin - S - oxide[8,9]
ring rupture

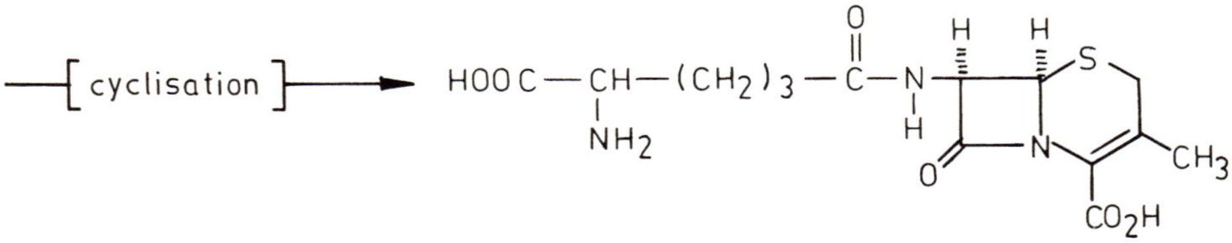

Deacetoxy cephalosporin C[10]

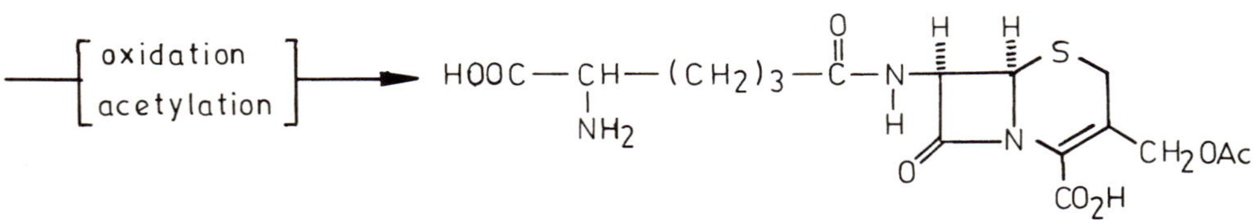

Cephalosporin-C

---

1. P. A. Lemke and D. R. Brannon in 'Cephalosporins and Penicillins', ed. E. H. Flynn, Academic Press, New York, 1972, p. 370-437.

2. 'The Chemistry of Penicillins', eds. H. T. Clarke, J. R. Johnson and R. Robinson, Princeton University Press, 1949, p. 861; J. C. Sheehan and K. R. Henery-Logan, J. Amer. Chem. Soc., 79, 1262 (1957); 81, 3089 (1959); E. P. Abraham, Quart. Rev., 21, 231 (1967); R. B. Woodward, K. Heusler, J. Gosteli, P. Naegeli, W. Oppolzer, R. Ramage, S. Ranganathan and H. Vorburggen, J. Am. Chem. Soc., 88, 852 (1966); R. B. Woodward, Science, 153, 487 (1966).

3. H. R. V. Arnstein and M. E. Clubb, Biochem. J., 65, 618 (1957); H. R. V. Arnstein and P. T. Grant, ibid., 57, 353, 360 (1954); S. C. Warren, G. G. F. Newton and E. P. Abraham, ibid., 103, 902 (1967).

4. The oxidation to thioaldehyde derivative can be rationalised as arising from rupture of a disulfide bond; the pattern of loss of hydrogen during oxidation has been confirmed by labelling studies:

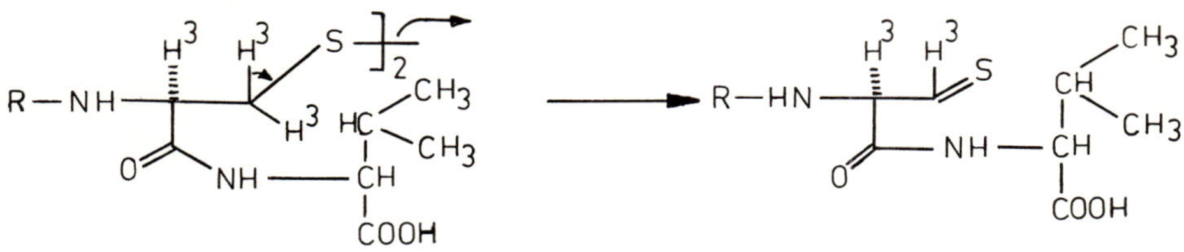

(H. R. V. Arnstein and J. C. Crawhall, Biochem. J., 67, 180 (1957)).

5. Extensive experimentation with labelled valine as summarised below,

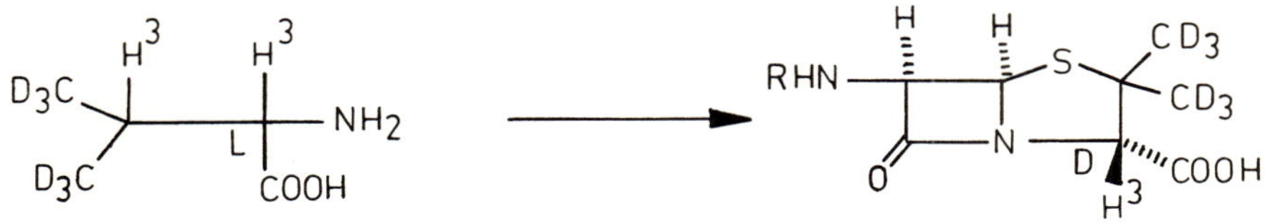

is in agreement with the dehydrovaline intermediate which is associated with the co-enzyme. It appears that in Michael addition that 'inverts' the C-2 centre, the $^3H_1$ that is lost during oxidation is picked up. (F. C. Huang, J. A. Chan, C. J. Sih, P. Fawcett and E. P. Abraham, J. Amer. Chem. Soc., 97, 3858 (1975); H. Kluender, F. C. Huang, A. Fritzberg, H. Schnoes, C. J. Sih, P. Fawcett and E. P. Abraham, J. Amer. Chem. Soc., 96, 4054 (1974)).

6. The in vitro Michael reaction of such a species is already known to yield penicillins (S. Wolfe, R. N. Bassett, S. M. Caldwell and F. I. Wasson, J. Amer. Chem. Soc., 91, 7205 (1969)).

7. Isopenicillin N has been found in the culture filtrates of *Penicillium chrysogenum* (E. H. Flynn, M. H. McCormick, M. C. Stamper, H. De. Valeria and C. W. Godzeski, J. Amer. Chem. Soc., 84, 4594 (1962); M. Cole and F. R. Batchelor, Nature, 198, 383 (1963)). Inversion at the α-amino group by a racemase gives penicillin N.

8. The ring expansion theory is supported by the fact that many organisms of the *Cephalosporium* type produce only penicillin N and mutants of *C. acremonium* have been isolated which are unable to synthesise cephalosporin-C but still retain the potential to synthesise penicillin N (P. A. Lemke and C. H. Nash, Can. J. Microbiol., 18, 255 (1972)).

9. The *in vitro* S-oxide → desacetoxy cephalosporin change has been accomplished:

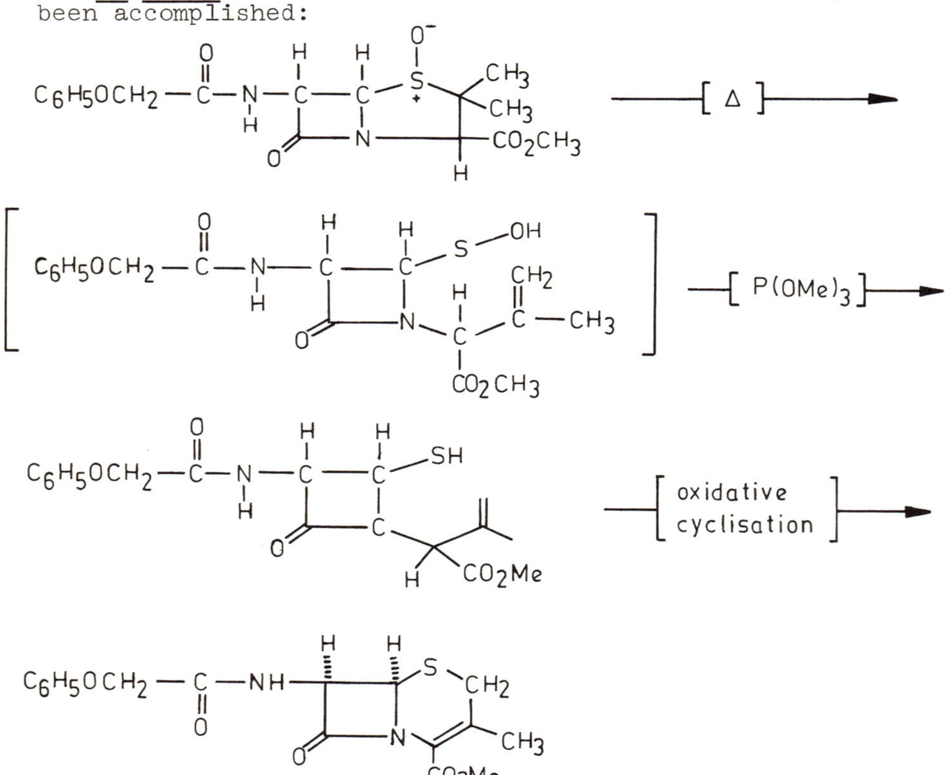

desacetoxy cephalosporin derivative

(R. B. Morin, B. G. Jackson, R. A. Mueller, E. R. Lavagnino, W. B. Scanlon and S. L. Andrews, J. Amer. Chem. Soc., 85, 1896 (1963); S. Wolfe, J. C. Godfrey, C. T. Haldrege and Y. G. Perron, J. Amer. Chem. Soc., 85, 643 (1963)).

10. Desacetoxycephalosporin C has been isolated from Streptomyces. (R. Nagarajan, L. D. Boeck, R. L. Hamill, C. E. Higgens and K. S. Yang, Chem. Comm., 321 (1974).)

11. An interesting variation involves the generation of the cepham and penam nucleus by concerted processes from a common precursor:

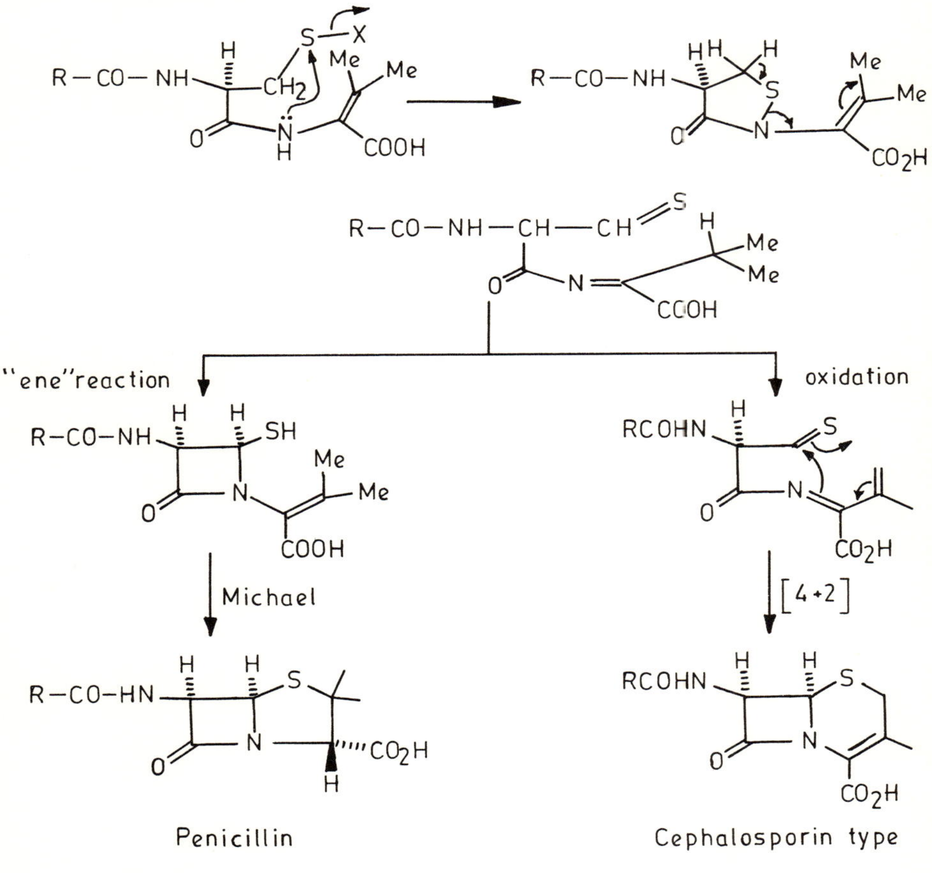

(J. E. Baldwin, S. B. Haber and J. Kitchin, Chem. Comm., 790 (1973).)

12. A biogenetic scheme, remarkably similar to Woodward's total synthesis of cephalosporin-C, starting from a locked cysteine has been proposed:

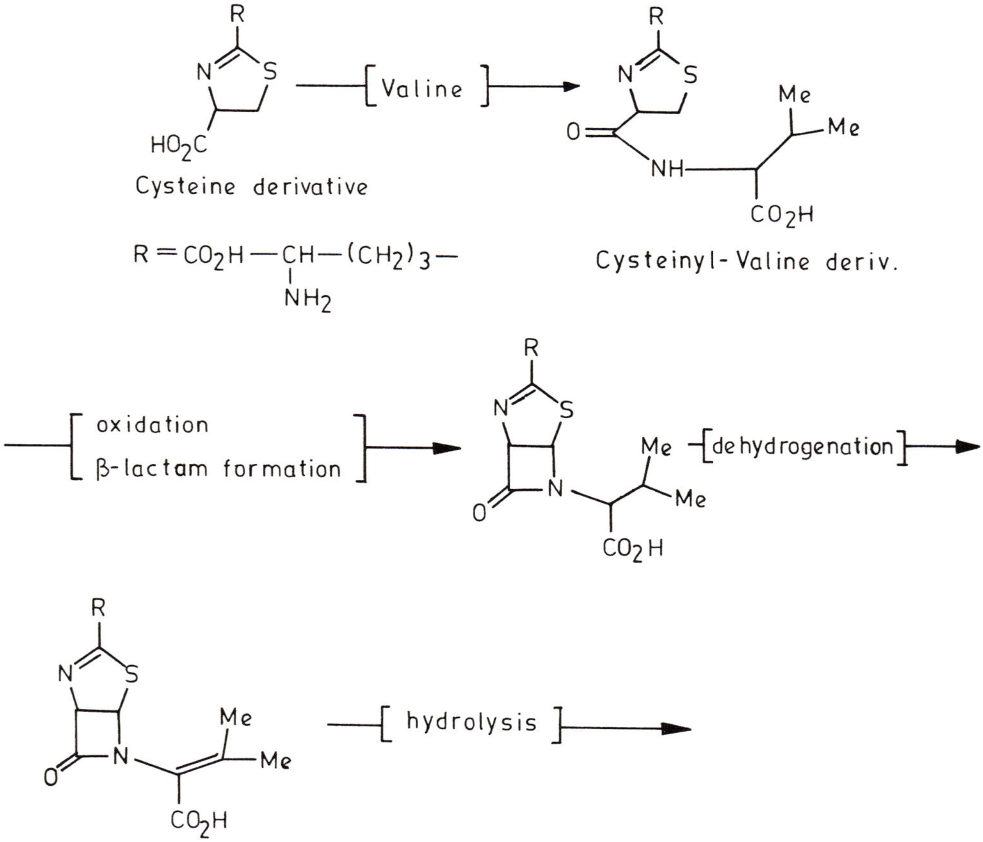

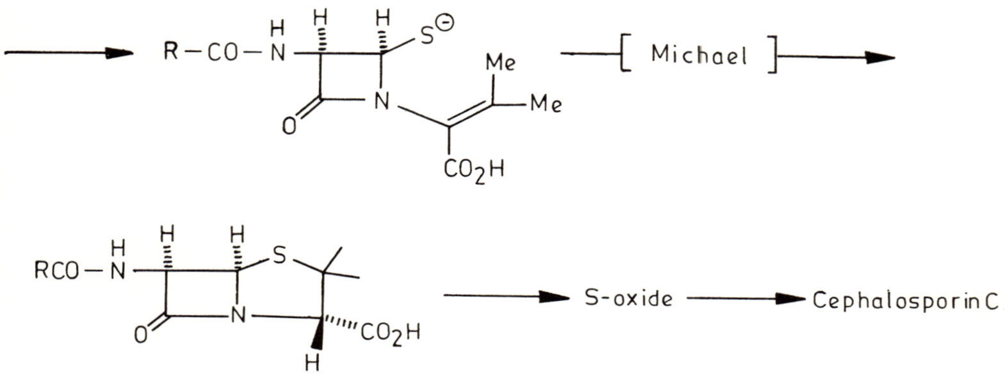

Penicillin

(R. D. G. Cooper, J. Amer. Chem. Soc., 94, 1018 (1972)).

A synthesis of penicillins and Cephalosporins based on the above suggestion has recently been accomplished. (S. Nakatsuka, H. Tanino and Y. Kishi, J. Amer. Chem. Soc., 97, 5008, 5010 (1975).)

CEPHALOTAXINE

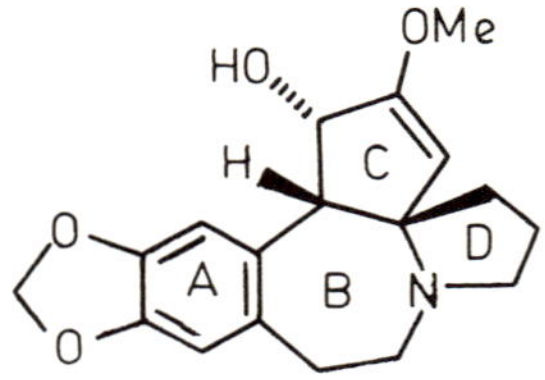

Cephalotaxine, the antileukemic alkaloid of the genus Cephalotaxus, is made from tyrosine in an unusual manner. The basic strategy of Nature[1] is to produce a key intermediate arising from an aromatic fission which is correctly disposed to generate the Cephalotaxine skeleton <u>via</u> a double cyclisation and rearrangement.

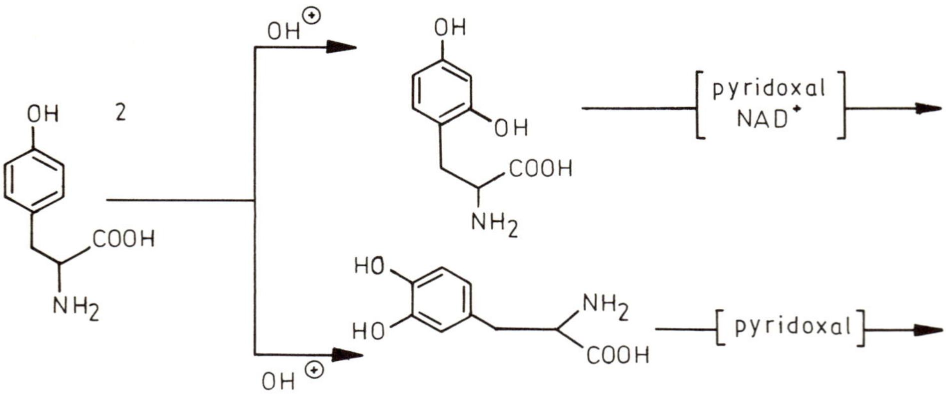

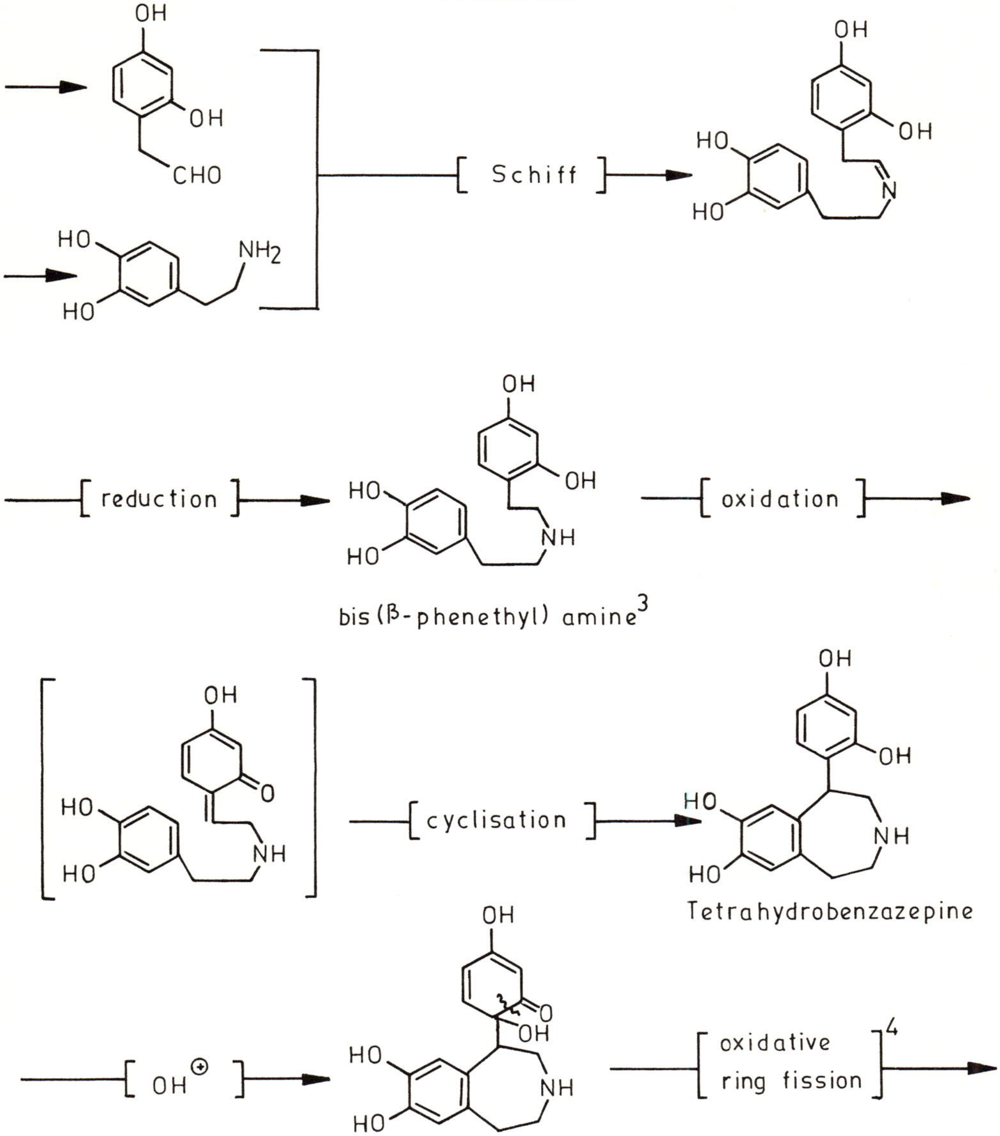

OH
OH
CHO
HO
HO
NH2
Schiff
OH
OH
HO
HO
N
reduction
OH
OH
HO
HO
NH
oxidation
bis (β-phenethyl) amine3
OH
O
HO
HO
NH
cyclisation
OH
OH
HO
HO
NH
Tetrahydrobenzazepine
OH
O
OH
HO
HO
NH
OH⊕
oxidative
ring fission
4

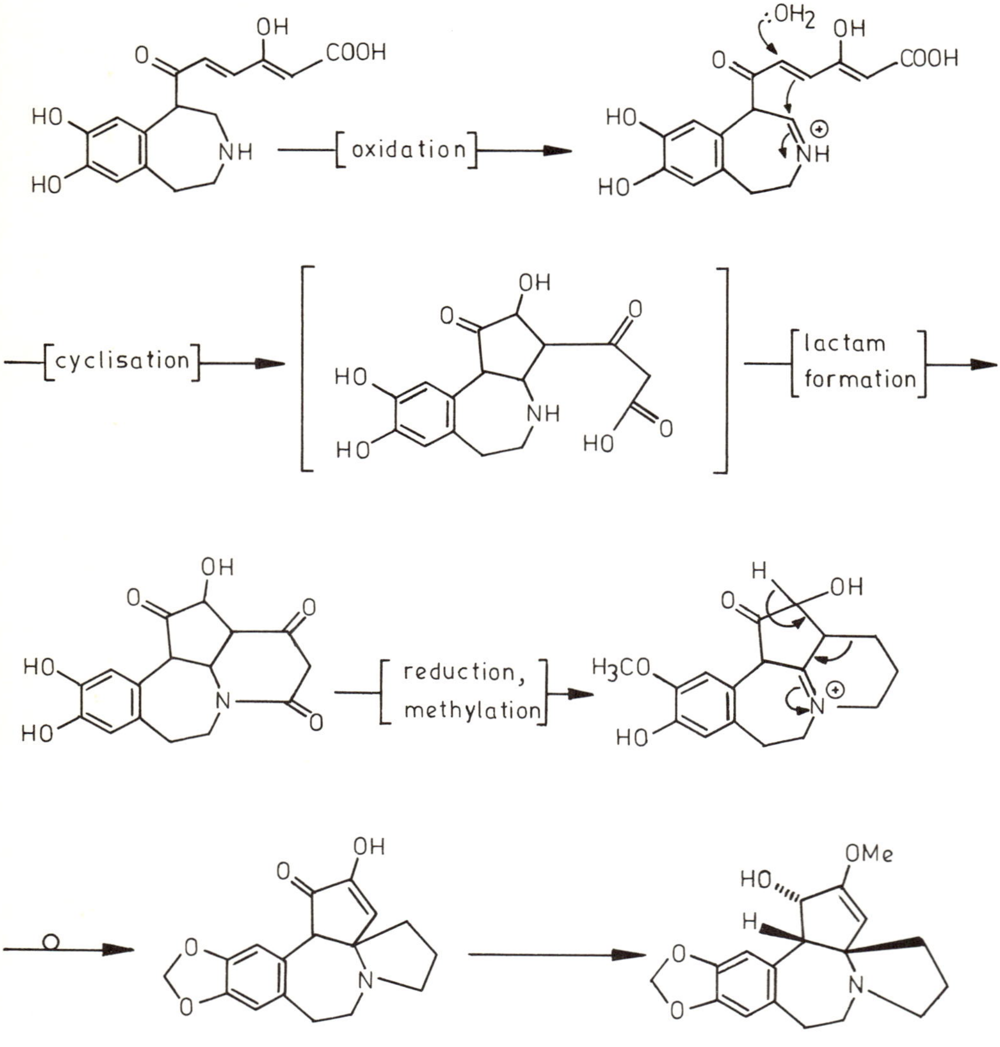

Desmethyl Cephalotaxinone[5]

Cephalotaxine

1. R. J. Parry and J. M. Schwab, J. Amer. Chem. Soc., 97, 2555 (1975).

2. Administration of [1-^{14}C], [2-^{14}C] and [3-^{14}C]-DL-tyrosine to young C. harringtonia plants yielded radioactive cephalotaxine (Ref. 1).

3. The mechanism of formation of bis(β-phenethyl) amine from tyrosine was suggested by Leete (E. Leete and A. Ahmad, J. Amer. Chem. Soc., 88, 4722 (1966)).

4. Oxidative fission of this aromatic ring is in a manner analogous to the degradation of aromatic rings observed in a variety of living systems (G. H. N. Towers and P. V. Subba Rao, Recent Adv. Phytochem., 4, 1 (1972)).

5. This intermediate has been isolated from Cephalotaxus plants (R. G. Powell and K. L. Mikolojczak, Phytochemistry, 12, 2987 (1973)).

CHLOROPHYLL

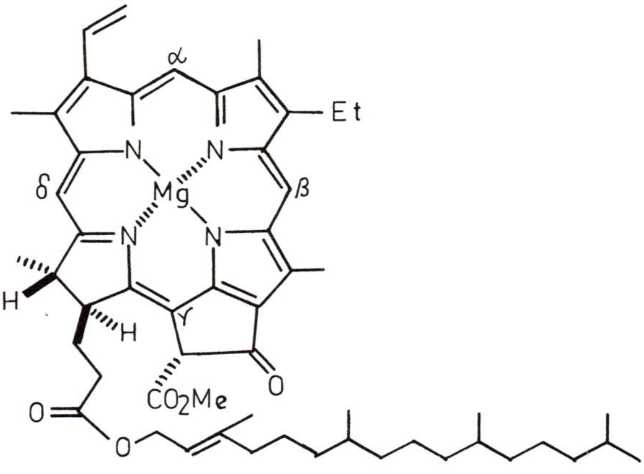

Nature, with its wily and amusing tricks, showing off the 'magic' that enzymes can bring about, and man, with a prodigious display of his synthetic armour, have both created chlorophyll, symbolyzing an adventure in the construction of molecular frameworks, which can, at best, be only equalled![1]

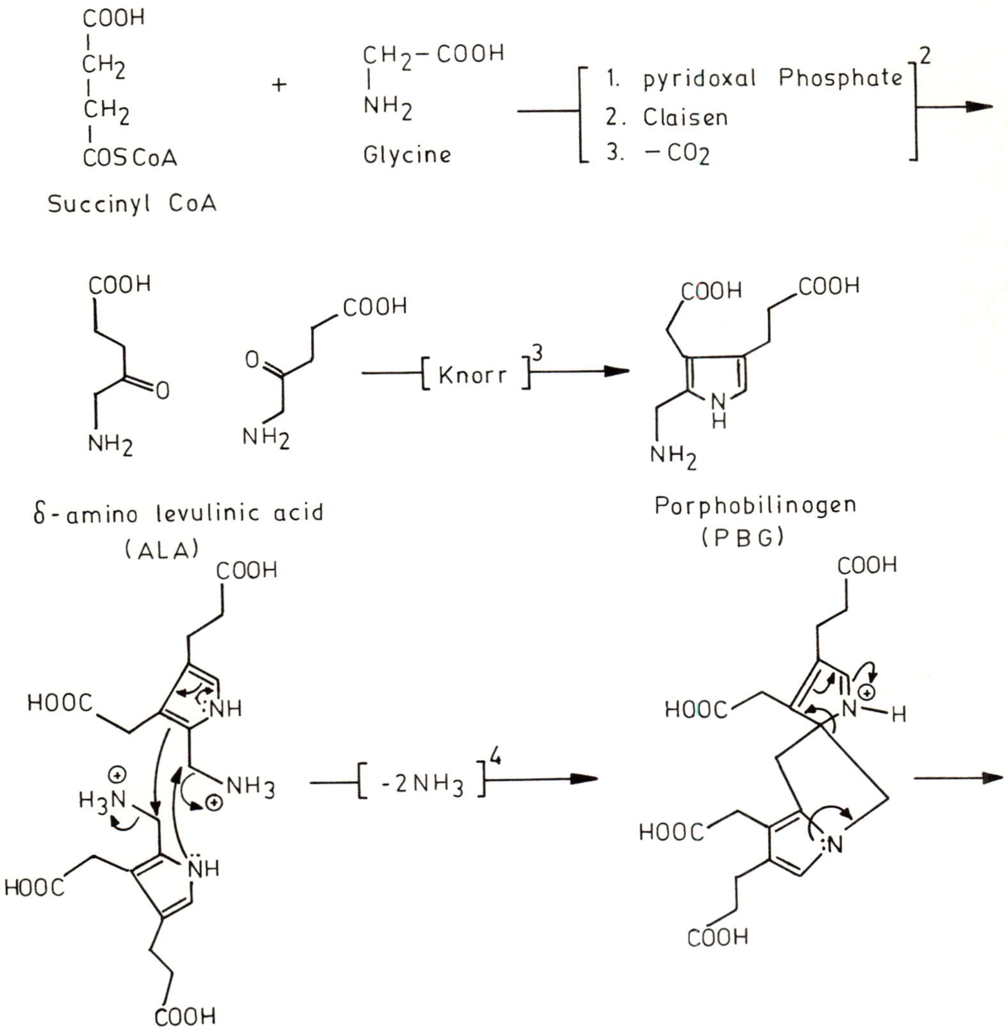
COOH
CH2
CH2
COSCoA
Succinyl CoA
+
CH2-COOH
NH2
Glycine
1. pyridoxal Phosphate
2. Claisen
3. - CO2
2
COOH
O
NH2
O
COOH
NH2
δ-amino levulinic acid
(ALA)
Knorr
3
COOH
COOH
N
H
NH2
Porphobilinogen
(PBG)
COOH
HOOC
NH
H3N
NH3
NH
HOOC
COOH
-2NH3
4
COOH
HOOC
N
H
HOOC
N
COOH

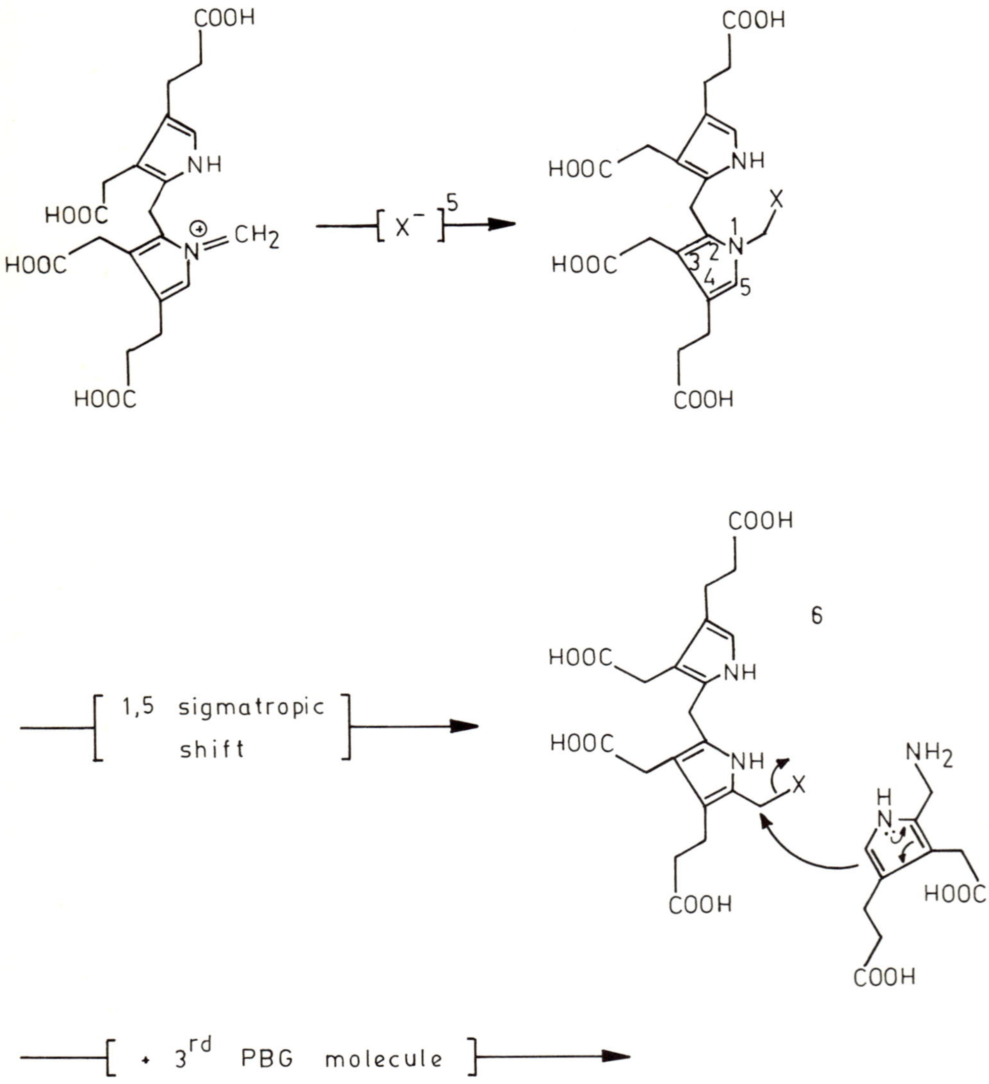
COOH
NH
HOOC
HOOC
CH2
HOOC
X⁻
5
COOH
HOOC
NH
X
1
HOOC
3
2
N
4
5
COOH
1,5 sigmatropic shift
COOH
6
HOOC
NH
HOOC
NH
X
NH2
H
N
HOOC
COOH
COOH
+ 3rd PBG molecule

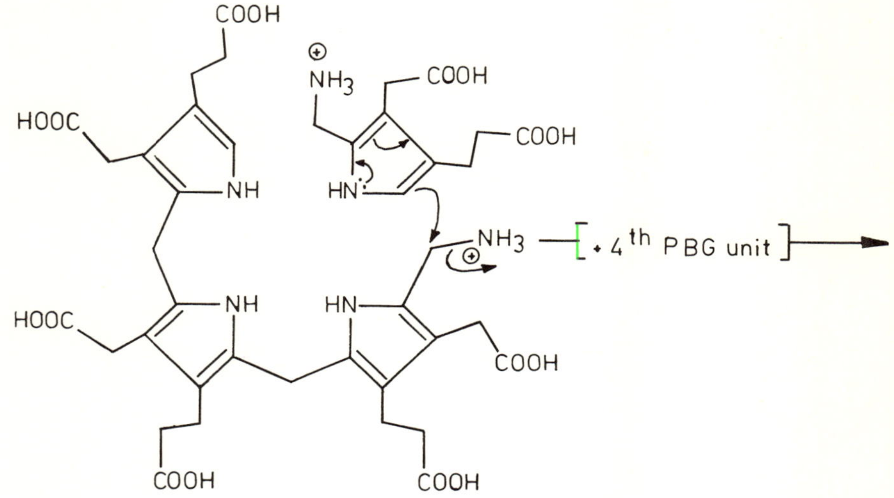
COOH
NH3
COOH
COOH
HOOC
NH
HN
NH3
[+ 4th PBG unit]
HOOC
NH
HN
COOH
COOH
COOH

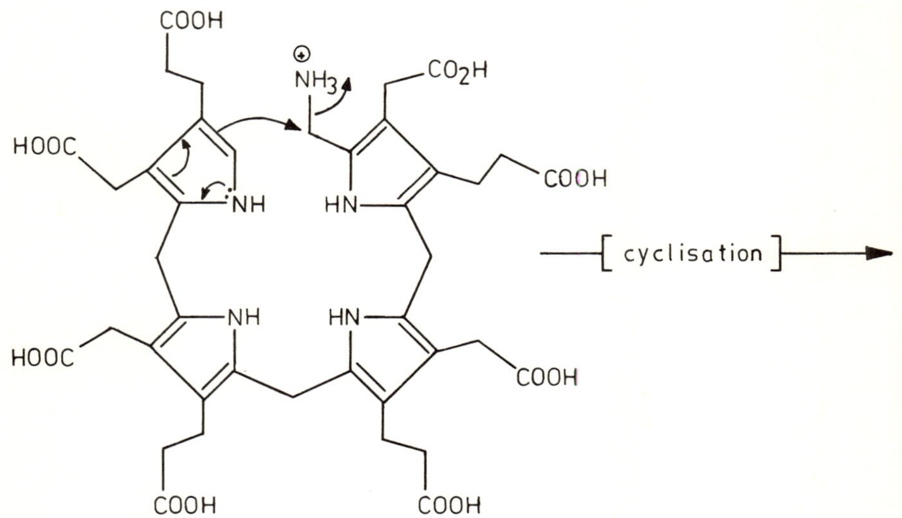
COOH
NH3
CO2H
HOOC
COOH
NH
HN
[cyclisation]
NH
HN
HOOC
COOH
COOH
COOH

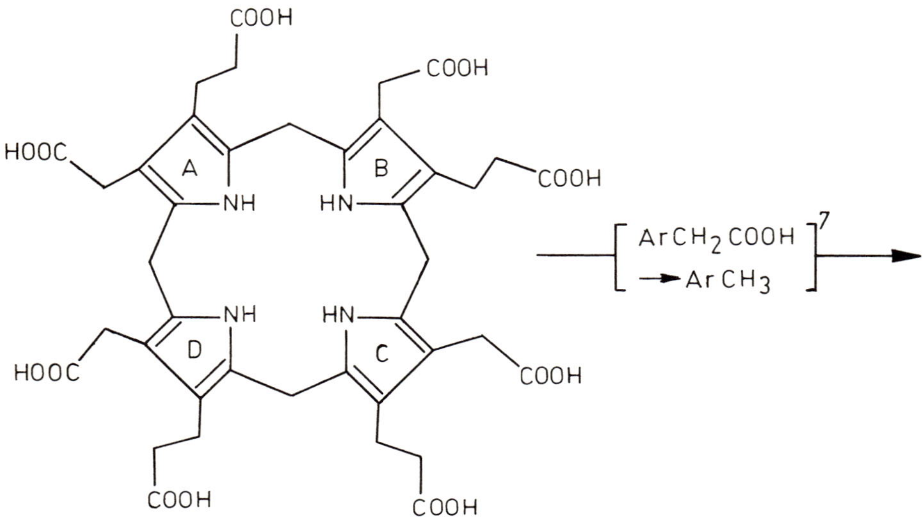

Uroporphyrinogen III

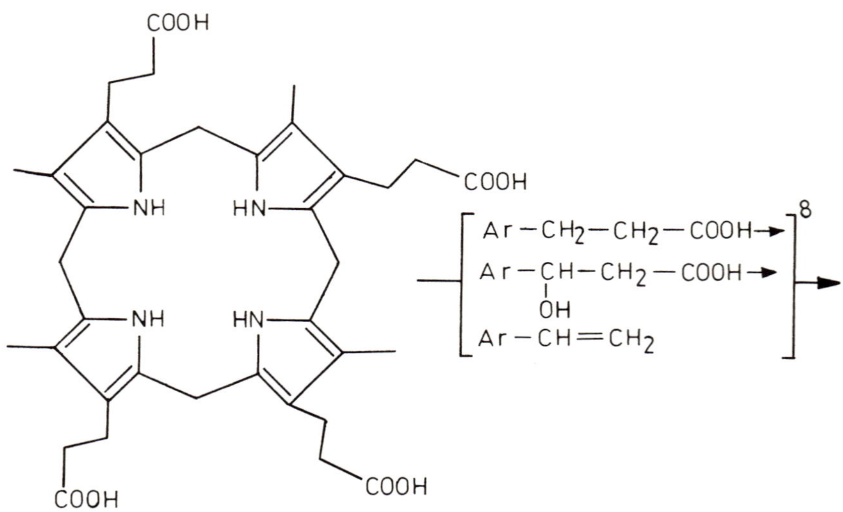

Coproporphyrinogen III

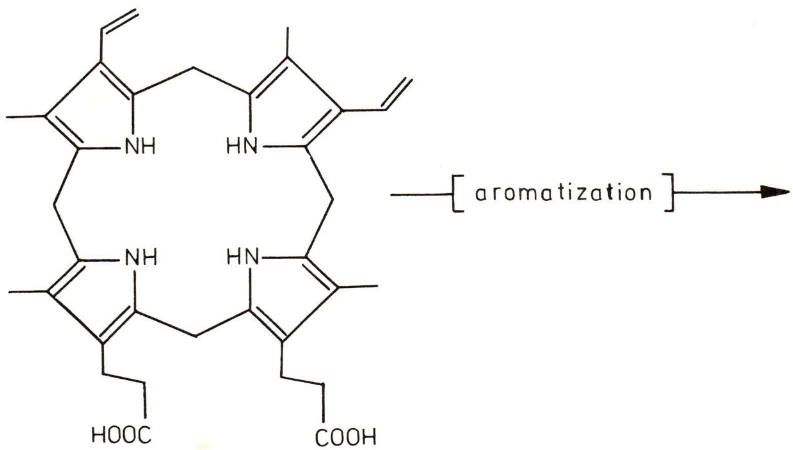

Protoporphyrinogen IX

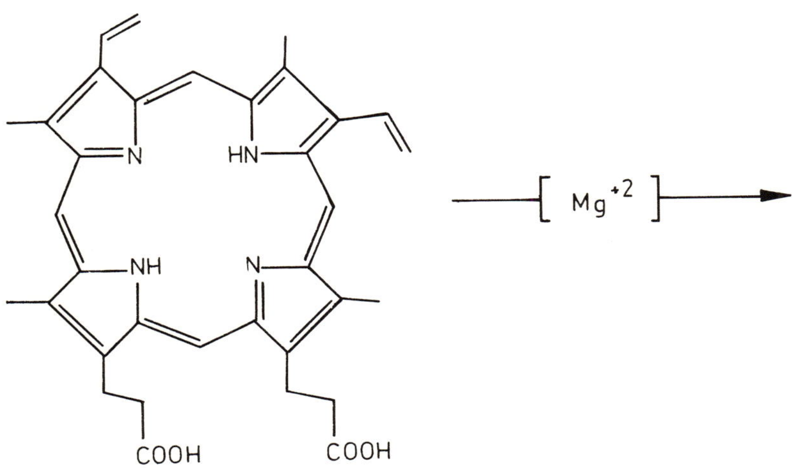

Protoporphyrin IX

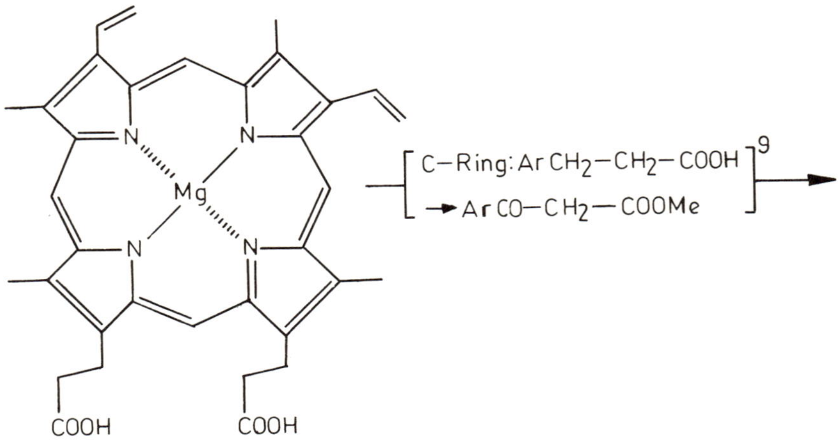

Mg– Protoporphyrin IX

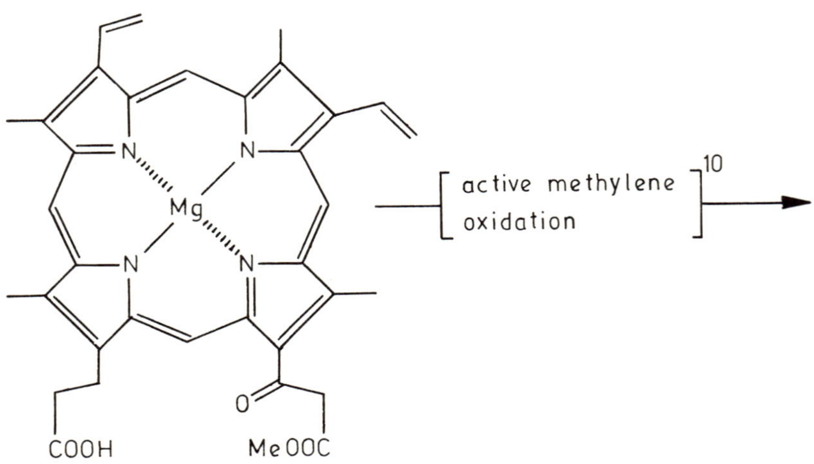

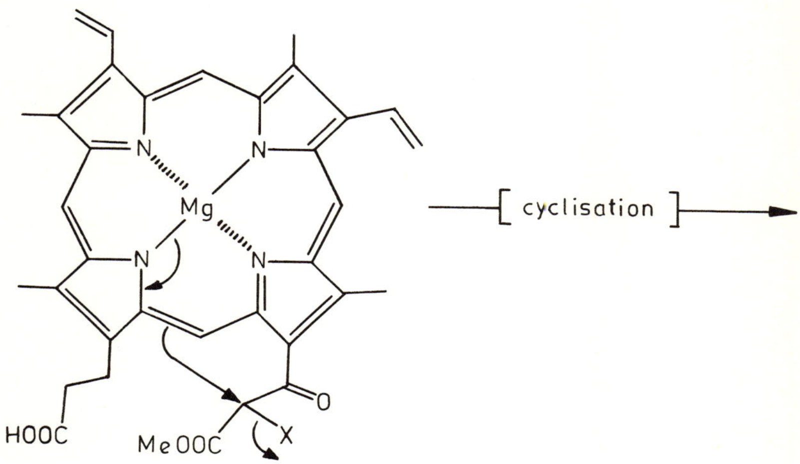
N
N
Mg
N
N
HOOC
MeOOC
O
X
cyclisation

11

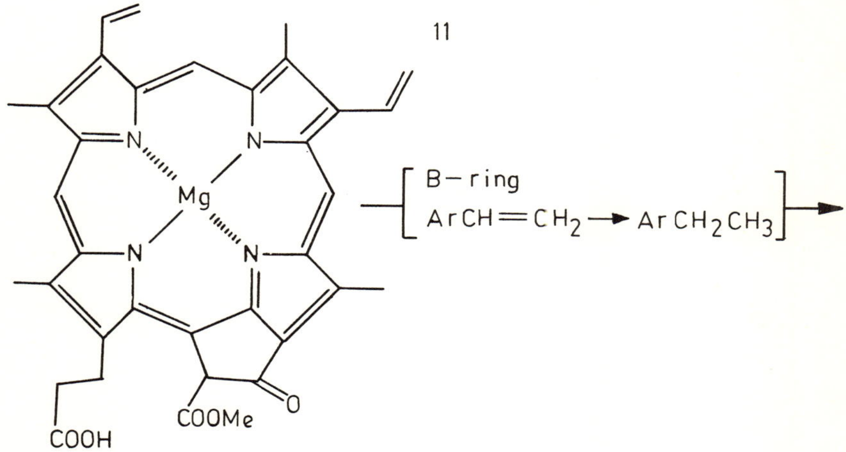
N
N
Mg
N
N
COOH
COOMe
O
B-ring
ArCH=CH2 → ArCH2CH3

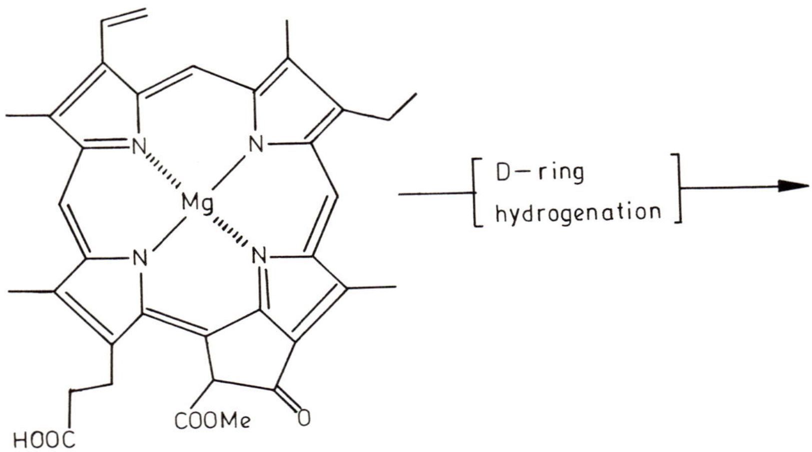

Protochlorophyllide[12]

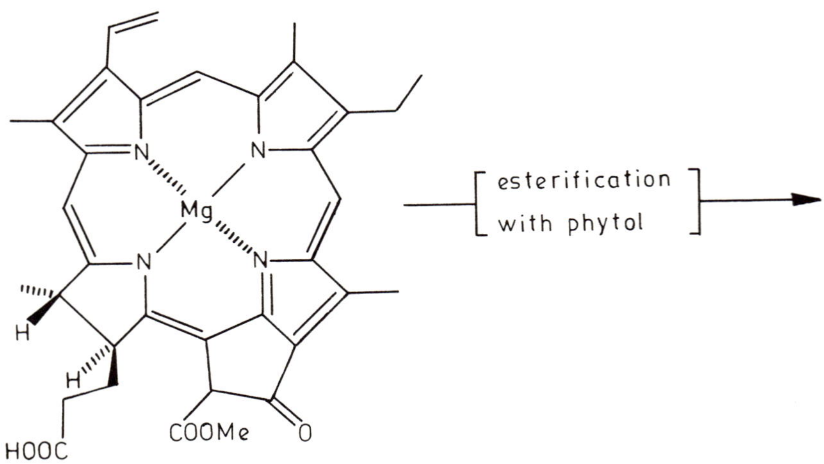

Chlorophyllide a[13]

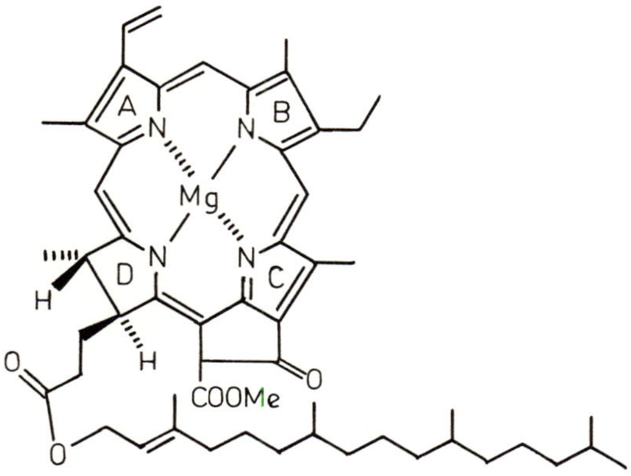

Chlorophyll a

---

1. L. Bogorad in 'The Chlorophylls', ed. L. P. Vernon and G. R. Seeley, Academic Press, New York, 1966, p. 481; B. F. Burnham in 'Metabolic Pathways' ed. D. M. Greenberg, Academic Press, New York, 1969, Vol. III, 3rd edn., p. 403; O. T. G. Jones in 'Porphyrins and Related Compounds' ed. T. W. Goodwin, Academic Press, New York, 1968, p. 131; O. T. G. Jones in 'Phytochemistry' ed. L. P. Miller, Van Nostrand Reinhold Co., New York, 1973, Vol. I, p. 75; For synthesis of chlorophyll see, R. B. Woodward et al., J. Amer. Chem. Soc., 82, 3800 (1960); R. B. Woodward, Angew. Chem., 72, 651 (1960).

2. The formation of ALA from succinyl CoA and glycine is the first step leading specifically to the biosynthesis of porphyrins and can be readily rationalised:

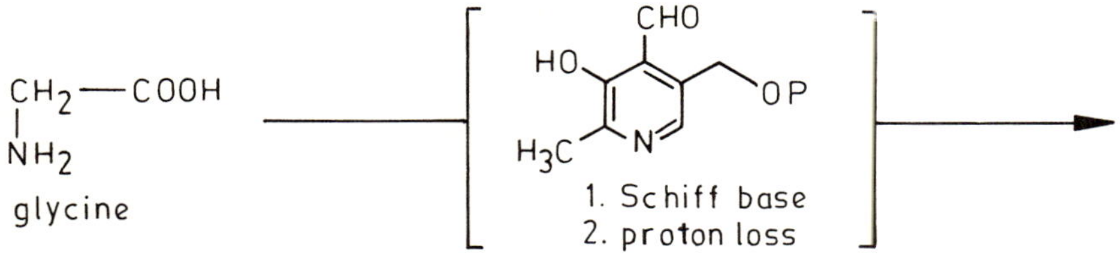

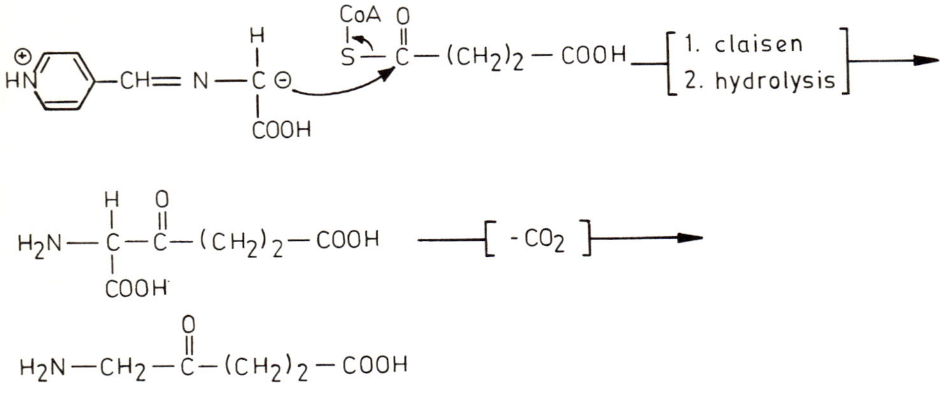

ALA

(A. Neuberger, Biochem. J., 78, 1 (1961)).

3. Careful studies with labelled substrates have shown that the first ALA is attached firmly to the enzyme by Schiff base formation. The second molecule of ALA is received at the enzyme active site in a conformation ideal for pyrrole formation:

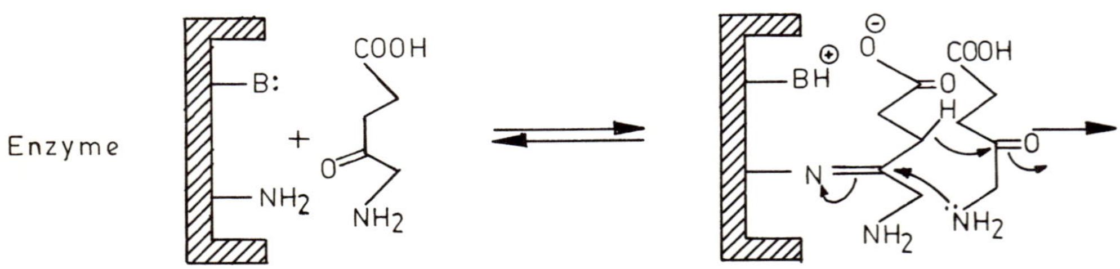

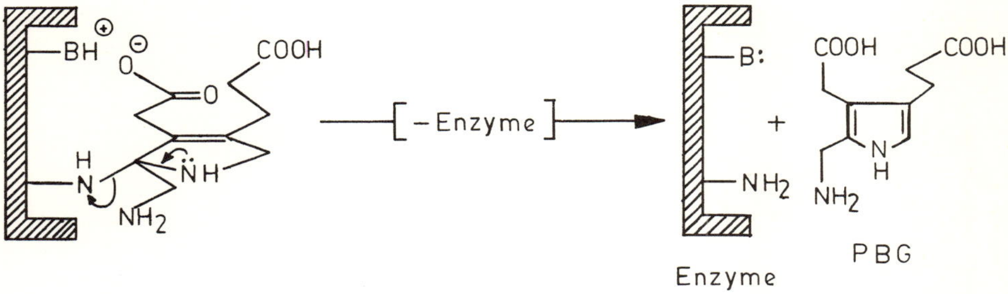

(D. Shemin, 'Mechanism and control of pyrrole synthesis' in "Porphyrins and Related Compounds" ed. T. W. Goodwin, Academic Press, New York, 1968, p. 75).

4. The conversion of 4PBG molecules into the cyclic tetrapyrrole uroporphyrinogen III takes place in the presence of PBG deaminase-urogen III cosynthetase complex. These two enzymes have been separated and isolated in pure form from wheat germ. The role of cosynthetase has been defined as a 'specific protein' of deaminase which changes the mode of PBG condensation from the beginning of the reaction and steering it towards the formation of urogen III.
More than 20 hypotheses have been proposed for the formation of uroporphyrinogen III from porphobilinogen (J. H. Mathewson and A. H. Corwin, J. Amer. Chem. Soc., 83, 135 (1961); A. R. Battersby, Int. Cong. Pure App. Chem., Spec. Lect. 23rd, 5, 1 (1971); E. Bullock, Nature (London), 205, 70 (1965); B. J. Whitlock, H. W. Whitlock and H. Alles, J. Amer. Chem. Soc., 96, 3959 (1974); P. Cornford, Biochem. J., 91, 64 (1964); J. Dalton and R. C. Dougherty, Nature (London), 223, 1151 (1969); E. B. C. Llambias and A. M. C. Battle, Biochem. J., 121, 327 (1971); R. Radmer and L. Bogorad, Biochemistry, 11, 904 (1972)). All these have recently been discarded in the light of new data obtained from the studies of chemical as well as enzymatic polymerisation of various synthetic 2-aminomethyldipyrryl methanes, 2-aminomethyltripyrranes and 2-aminomethyltetrapyrranes (bilanes) and that no free formaldehyde is liberated during biosynthesis (B. Frydman and R. B. Frydman, Acc. Chem. Res.,

8, 201 (1975)).
A rationalisation that takes into account all the experimental evidence envisages the initial union of two porphobilinogen units leading to a N-substituted dipyrrylmethane, which undergoes a facile 1,5 sigmatropic shift leading to a dipyrryl unit properly equipped for the formation of uroporphyrinogen III by accepting additional 2 units of porphobilinogen in an uncomplicated manner.

5. In the in vivo system X^- is a donor that is present at the active site of the enzyme.

6. This dipyrrylmethane was found to be incorporated into urogen III. This incorporation agrees with results obtained by studying the incorporation of double labelled [2,11-^{13}C] PBG into protoporphyrin IX (A. R. Battersby, E. Hunt and E. McDonald, J. Chem. Soc., Chem. Comm., 442 (1973)).

7. The decarboxylation of the four acetic acid side chains to four methyl groups occurs one at a time. However, the order of acetic acid chains are not known (L. C. Viale and M. Grinstein, Biochim. biophys. Acta., 158, 79 (1968)).

8. This transformation can be readily understood in terms of a decarboxylative elimination:

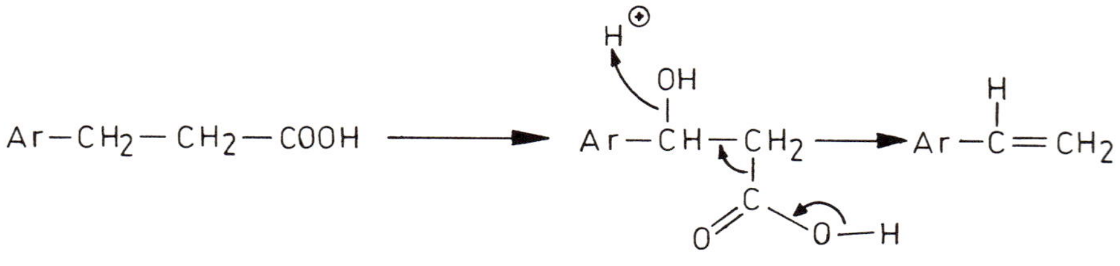

(S. Sano, J. biol. chem., 241, 5276 (1966)).

9. Magnesium protoporphyrin monomethyl ester is found in some chlorophyll deficient mutants of chlorella (S. Granick, J. biol. chem., 236, 1168 (1961)) and wild type R. spheroides (O. T. G. Jones, Biochem. J., 86, 429 (1963)). The enzyme

system that carries out the methylation of magnesium protoporphyrin was characterised as S-adenosyl methionine-magnesium protoporphyrin methyl transferase; the methyl group donor was found to be S-adenosyl methionine (K. D. Gibson, A. Neuberger and G. H. Tait, Biochem. J., 88, 483 (1963)). Subsequent to methylation the Ar-CH_2- ⟶ Ar-C(=O)- change takes place (L. Bogorad in 'The Chlorophylls' eds. L. P. Vernon and G. R. Seely, Academic Press, New York, 1966, p. 481).

10. The further transformation of the β-keto ester system to the cyclised product could be readily understood in terms of prior oxidation of the highly reactive methylene.

11. There is evidence that this pigment is formed by photosynthetic organisms (O. T. G. Jones, Biochem. J., 88, 335 (1963)).

12. A mutant of chlorella that accumulates protochlorophyllide has been isolated (S. Granick, J. biol. chem., 183, 713 (1950)).

13. Each molecule of protochlorophyllide is associated with a specific protein - the halochrome. This complex on illumination, even at -60°, leads to the formation of the chlorophyllide-a halochrome (N. K. Boardman, Biochim. biophys. Acta, 62, 63 (1962); J. H. C. Smith in 'Comparative Biochemistry of Photoreactive Systems' ed. M. B. Allen, Academic Press, New York, 1960, p. 257).

CHOLESTEROL

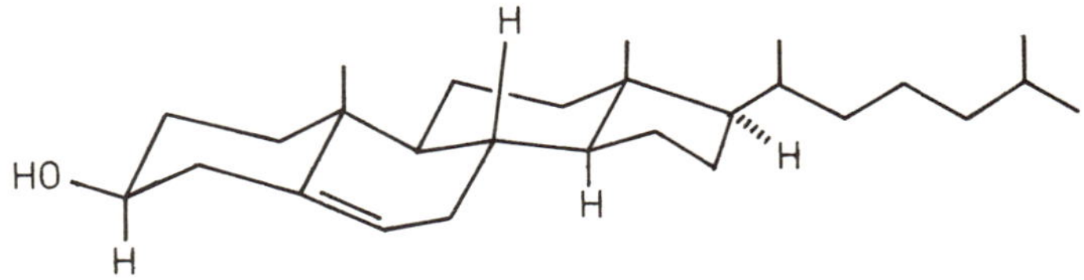

Although the intermediacy of squalene in cholesterol biosynthesis was conceived as early as 1926,[1] it was not until 1953 that this assumption was firmly confirmed by impressive and extensive experimentation.[2] The chemical simulation of the squalene cyclization has been the object of intense scrutiny in the recent past.[3]

[4]

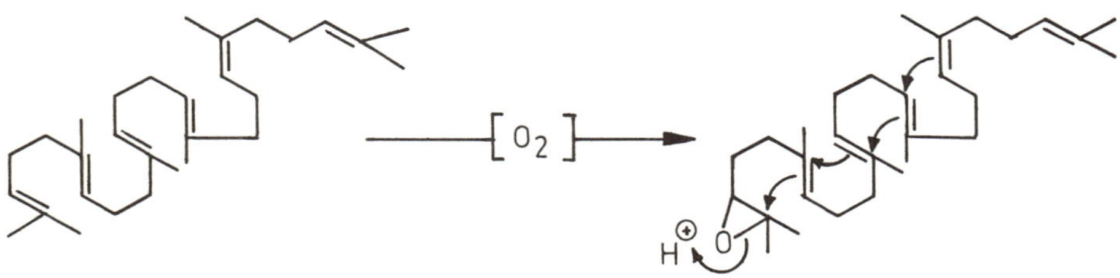

Squalene — 2,3-oxidosqualene

[5,6]

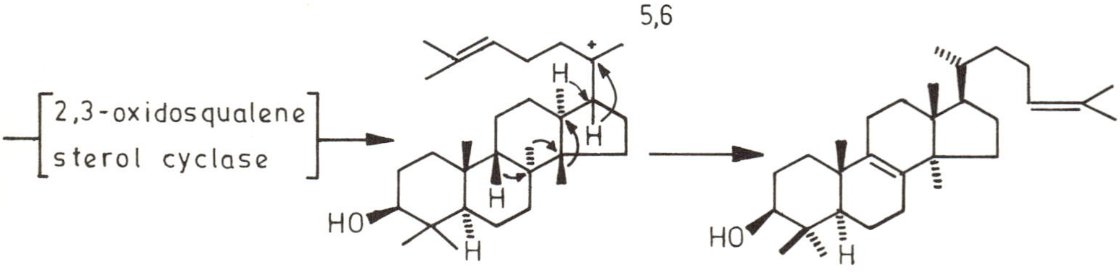

Protolanosterol — Lanosterol

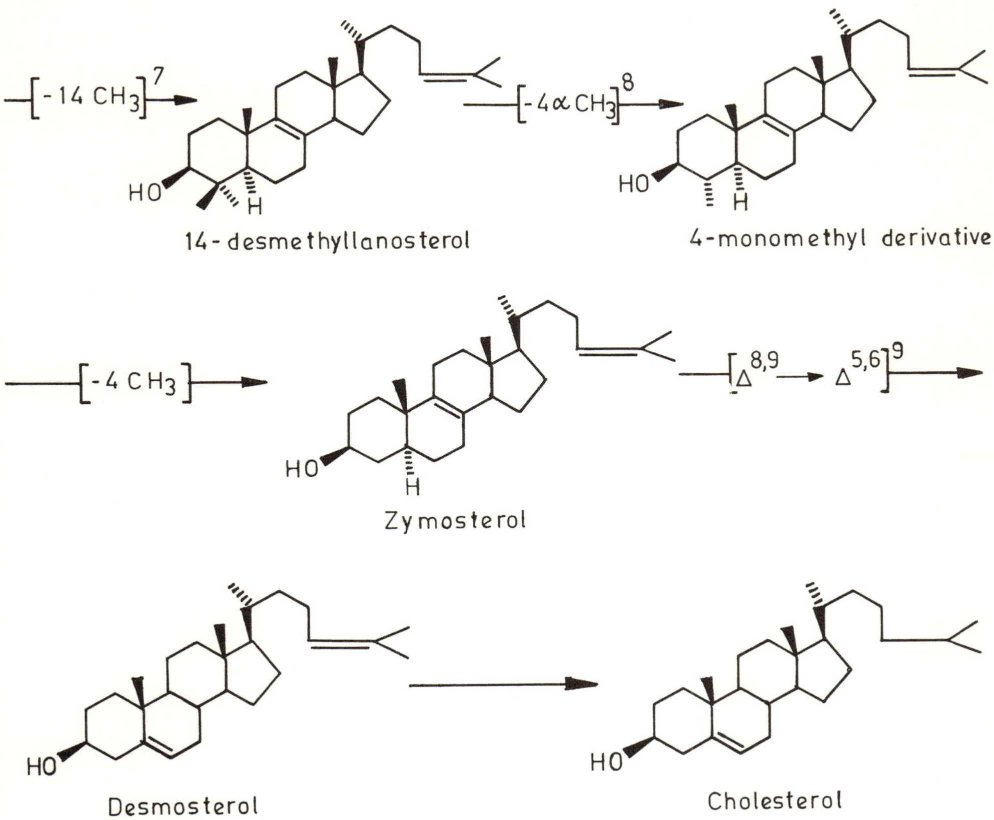

---

1. H. J. Channon, Biochem. J , 20, 400 (1926).

2. The discovery of the intermediates and the stereochemistry of the enzymatic processes by which squalene is biosynthesised from mevalonic acid has been described. R. B. Clayton, Quart. Rev., 19, 168 (1965); J. W. Cornforth and G. Popjak, Biochem. J., 101, 553 (1966); J. W. Cornforth, Quart. Rev., 23, 125 (1969); L. J. Mulheirn and P. J. Ramm, Chem. Soc. Rev., 1, 259 (1972).

3. W. S. Johnson; Acc. Chem. Res., 1 1 (1968); E. E. van Tamelen, *ibid*., 1, 111 (1968).

4. Intermediacy of 2,3-oxidosqualene in lanosterol biosynthesis in rat liver homogenates has been shown on the basis of experimental evidence; (E. J. Corey, W. E. Russey and P. R. Ortiz de Montellano, J. Amer. Chem. Soc., 88, 4750 (1966); E. J. Corey, P. R. Ortiz de Montellano, K. Lin and P. D. G. Dean, J. Amer. Chem. Soc., 89, 2797 (1967); E. E. van Tamelen, J. D. Willett, R. B. Clayton, and K. E. Lord, J. Amer. Chem. Soc., 88, 4752 (1966); E. E. van Tamelen, J. D. Willett and R. B. Clayton, J. Amer. Chem Soc., 89, 3371 (1967).

5. The formation of this intermediate from squalene epoxide requires the generation of a comparatively unstable ring B boat in intermediate generated by chair-boat-chair cyclisation.

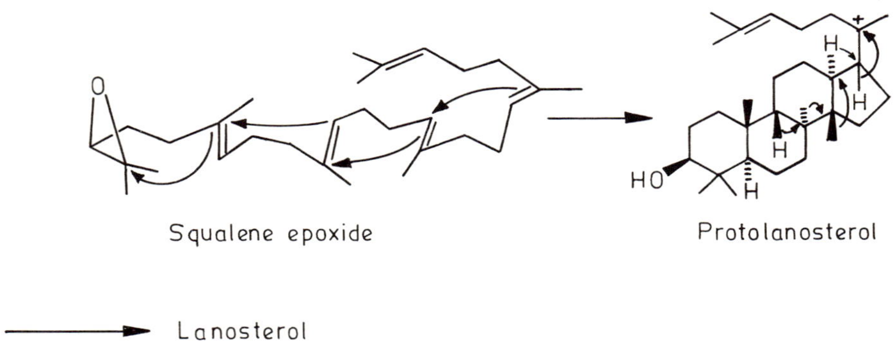

6. Protolanosterol, having, in addition to 8α- and 14β-methyls, the unusual 9β,10β(*cis*) relationship of hydrogen and methyl, is logically set for formation of lanosterol structure and stereochemistry by means of a series of 1,2-methyl,hydrogen shifts and C-9 proton loss (E. E. van Tamelen, J. D. Willett, R. B. Clayton, and K. E. Lord, J. Amer. Chem. Soc., 88, 4752 (1966).

7. The initial experiments of Bloch and coworkers (J. A. Olson, M. Lindberg and K. Bloch, J. Biol. Chem., 226, 941 (1957); F. Gautschi and K. Bloch, J. Amer. Chem. Soc., 79, 684 (1957); J. Biol. Chem., 233, 1343 (1958)) suggested that C-14

demethylation involves oxidation to the carboxylic acid followed by loss of CO_2 and this process precedes those at C-4. This was supported by the observation that 14α-hydroxymethyl and 14α-formyl derivatives of 4,4-dimethylcholest-7-en-3β-ol are metabolised to cholesterol by liver microsomes (J. Fried, A. Dudowitz and J. W. Brown, Biochem. Biophys. Res. Comm., 32, 568 (1968). Recently it has been demonstrated by using [32-H^3]lanost-7-en-32-3β-diol with microsomal enzyme system that C-14 demethylation occurs not by decarboxylation but by release of formic acid at the aldehyde level of oxidation (K. Alexander, M. Akhtar, R. B. Boar, J. F. McGhie and D. H. R. Barton, Chem. Comm., 383 (1972)). This C-14 demethylation, could, in principle, be chemically realized by initial functionalization of the homoallylic methyl group.

8. The 4α-methyl is eliminated first. Subsequently the remaining methyl takes up the 4α-orientation and is eliminated by a similar sequence:

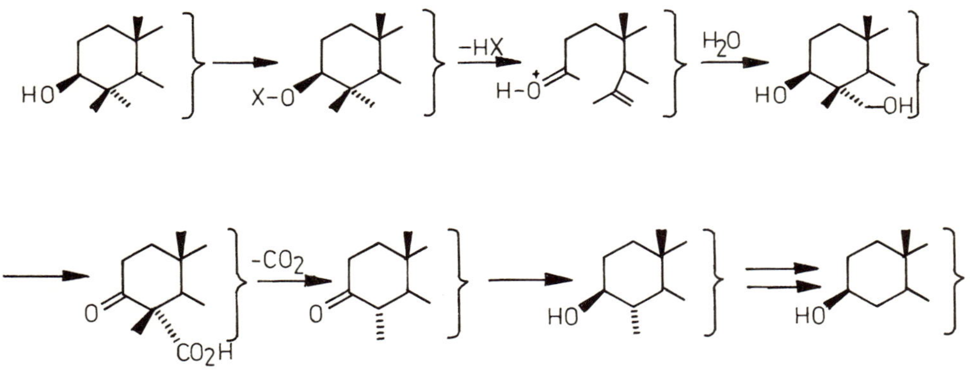

(J. L. Gaylor and H. S. Mason, J. Biol. Chem., 243, 4966 (1968); A. C. Swindell and J. L. Gaylor, ibid., p. 5546; N. J. Moir, W. L. Miller, and J. L. Gaylor, Biochem. Biophys. Res. Comm., 33, 916 (1968); W. L. Miller and J. L. Gaylor, J. Biol. Chem., 245, 5369, 5375 (1970); G. M. Hornby and G. S. Boyd, Biochem. Biophys. Res. Comm., 40, 1452 (1970); K. B. Sharpless, T. E. Snyder, T. A. Spencer, K. K. Maheshwari, J. A. Nelson, and R. B. Clayton, J. Amer. Chem. Soc., 91, 3394 (1969)).

9. The transposition of the Δ^8 double bond has been demonstrated to take place via Δ^7 ene and $\Delta^{5,7}$ diene; M. Akhtar, Biochem. J., 114, 71 (1969); ibid., 102, 462 (1967); M. E. Dempsey, J. Biol. Chem., 240, 4176 (1965).

COLCHICINE

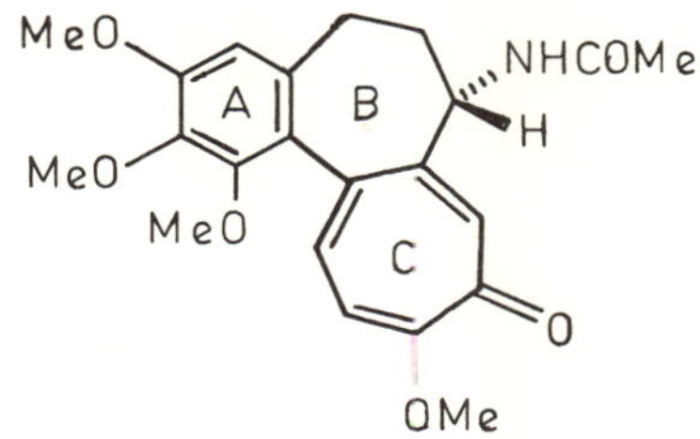

The heart of colchicine is the tropolone system whose presence was predicted in one of the most amazing intuitive feats! - even before the unit was characterized![1] It is fascinating to note that nature makes the tropolone unit and colchicine[2] in a manner quite divergent to that accomplished by six research groups.[3]

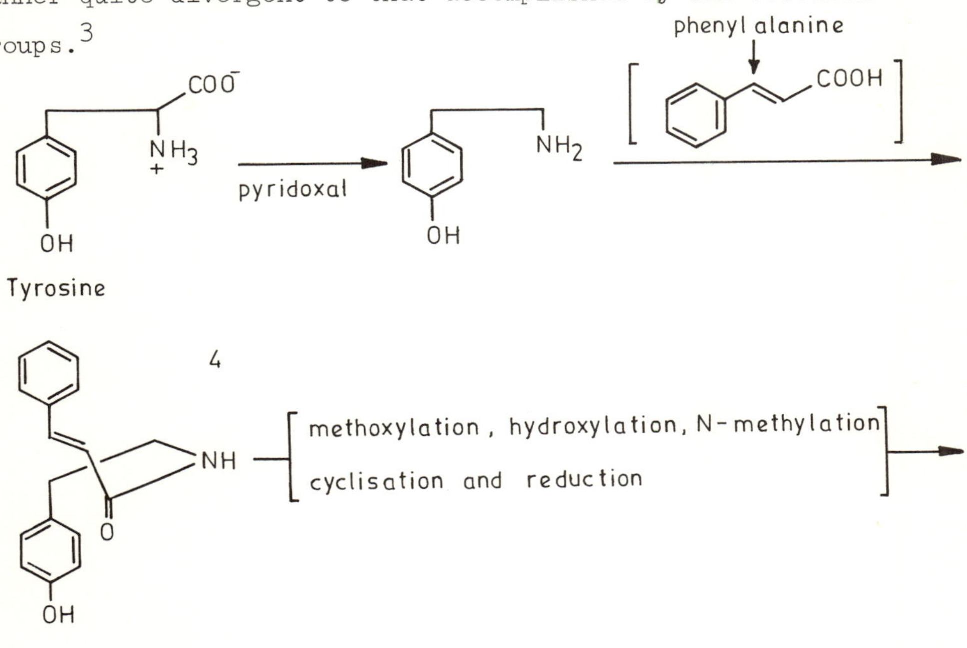

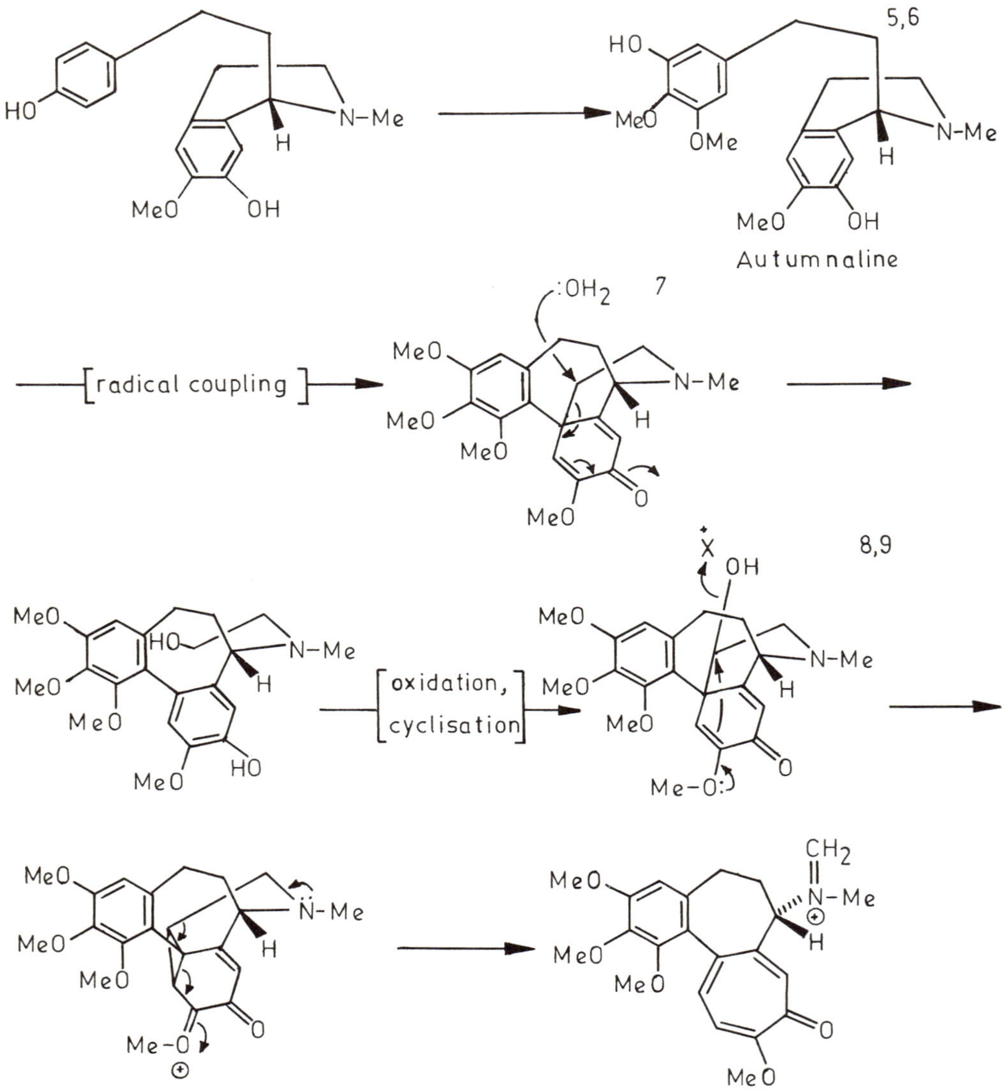
5,6
Autumnaline
radical coupling
7
8,9
oxidation,
cyclisation

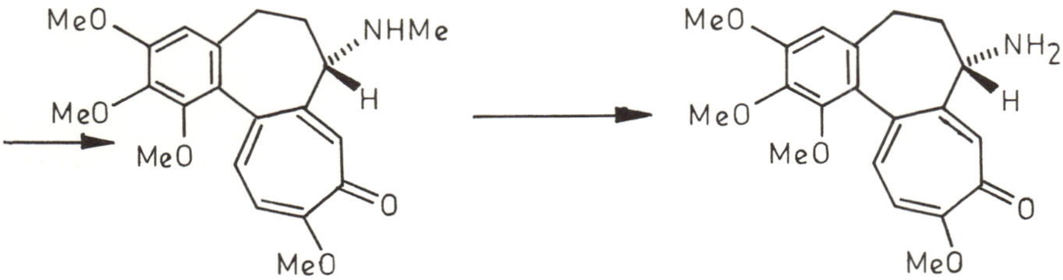

Demecolcine

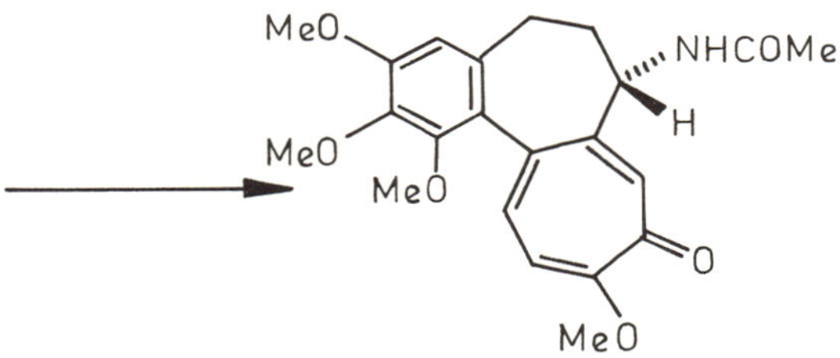

Colchicine

---

1. M. J. S. Dewar, Nature, 155, 50 (1945).

2. A. R. Battersby and J. J. Reynolds, Proc. Chem. Soc., 346 (1960); A. R. Battersby, R. Binks and D. A. Yeowell, ibid., 86 (1964); A. R. Battersby, R. Binks, J. J. Reynolds and D. A. Yeowell, J. Chem. Soc., 4257 (1964); A. R. Battersby, R. B. Herbert, E. McDonald, R. Ramage and J. H. Clements, J. Chem. Soc., Perkin I, 1741 (1972); E. Leete and P. E. Nemeth, J. Amer. Chem. Soc., 82, 6055 (1960); E. Leete, ibid., 85, 3666 (1963).

3. (a) J. Schreiber, W. Leimgruber, M. Pesaro, P. Schudel and A. Eschenmoser, Angew. Chem., 71, 637 (1959); Helv. Chim. Acta, 44, 540 (1961); (b) E. E. van Tamelen, T.A. Spencer, D. S. Allen and R.L. Orvis, J. Amer. Chem. Soc., 81, 6341 (1959); Tetrahedron, 14, 8 (1961); (c) R. B. Woodward, 'Harvey Lectures Series', 59, 31 (1963-64); (d) A. I. Scott, F. McCapra, R. L. Buchnan, A. C. Day and D. W. Young, Tetrahedron, 21, 3605 (1965); (e) T. Nakamura, Chem. Pharm. Bull., 8, 843 (1960); S. Sunagawa, J. Nakamura and K. Nakazawa, Chem. Pharm. Bull., 9, 81 (1961); 10, 281 (1962);

(f) E. Kotani, F. Miyazaki and S. Tobinaga, Chem. Comm., 300 (1974).

4. This compound has recently been shown to be an efficient precursor of the tropolone alkaloids in *colchicum autumnale* plant (A. R. Battersby, R. B. Herbert, E. McDonald, R. Ramage and J. H. Clements, J. Chem. Soc. Perkin I, 1741 (1972).

5. Satisfactory incorporations into demecolcine and colchicine were achieved when automnaline was fed to *C-byzantinum* plants. This supported the biosynthetic relationship between colchicine and 1-phenethylisoquinoline system. Recently[1-^{13}C]automnaline was incorporated successfully into colchicine in *C-autumnale* plants; (A. R. Battersby, P. W. Sheldrake and J. A. Milner, Tetrahedron Letters, 3315 (1974).

6. The oxidative coupling step, autumnaline→O-methylandrocymbine produces a new C-C bond between the positions *para* to the phenolic -OH groups. Direct tracer study of such C-C couplings has consistently shown that phenolic hydroxyl groups *ortho* or *para* to the new bond are essential. (A. R. Battersby in 'Oxidative Coupling of Phenols', eds. W. I. Taylor and A. R. Battersby, Dekker, New York, 1967, p. 119; D. H. R. Barton and T. Cohen, 'Festchr. A. Stoll', Birkhauser, Basle, 1957, p. 117).

7. Administration of Labelled O-methylandrocymbine to *C-autumnale* plants produced high incorporation, thereby clearly defining the biosynthetic path as autumnaline O-methylandrocymbine → colchicine (see ref. 5).

8. Recent work of Battersby (A. R. Battersby, T. A. Dobson, D.M. Foulkes and R. B. Herbert, J. Chem. Soc., Perkin I, 1730 (1972)) has shown that when tyrosine[3-^{14}C] was fed to *colchicum autumnale* plants, the tyrosine label was specifically located at C-12 in the tropolone ring thereby establishing that tropolone system is generated from the aromatic nucleus of tyrosine by a ring expansion process with inclusion of the benzylic carbon atom.

9. This is similar to the non-enzymatic case i→ii:

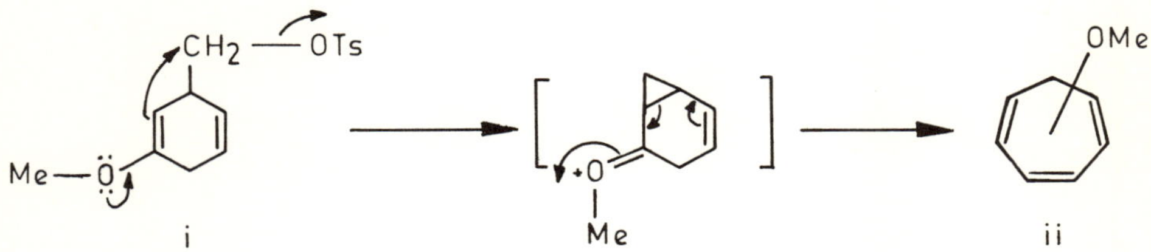

i   ii

(O. L. Chapman and P. Fitton, J. Amer. Chem. Soc., 85, 41 (1963)).

ERYTHRALINE

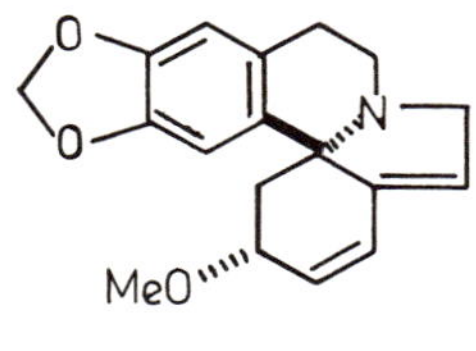

A fascinating spiro→spiro change is the central strategy in the biosynthesis[1] of Erythraline, the representative of the Erythrina group of alkaloids.

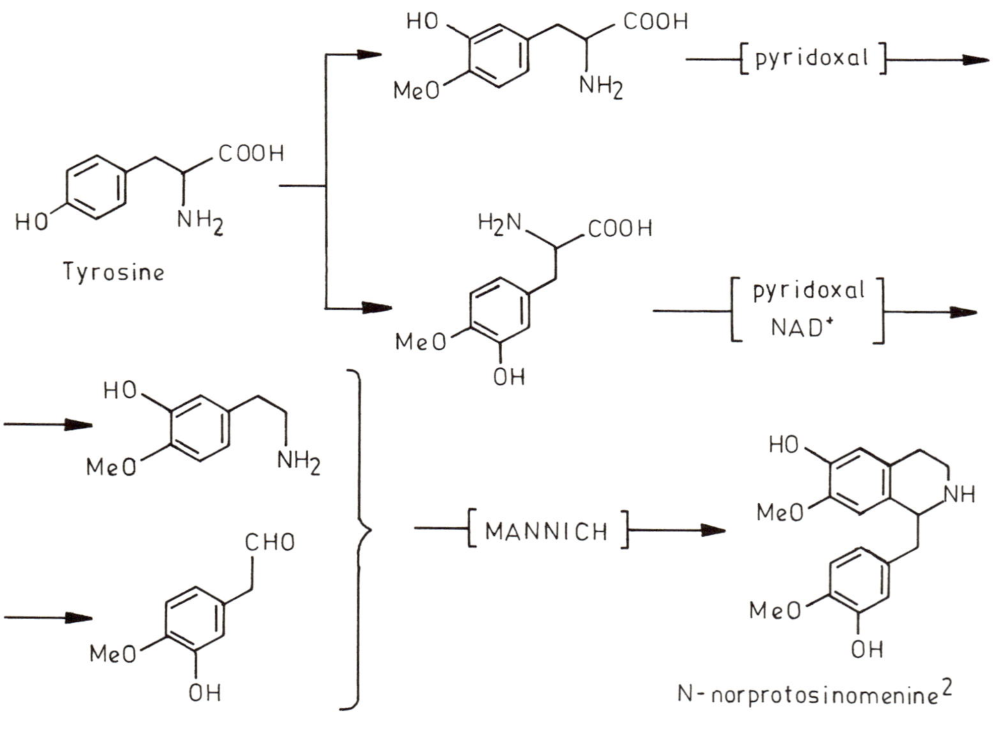

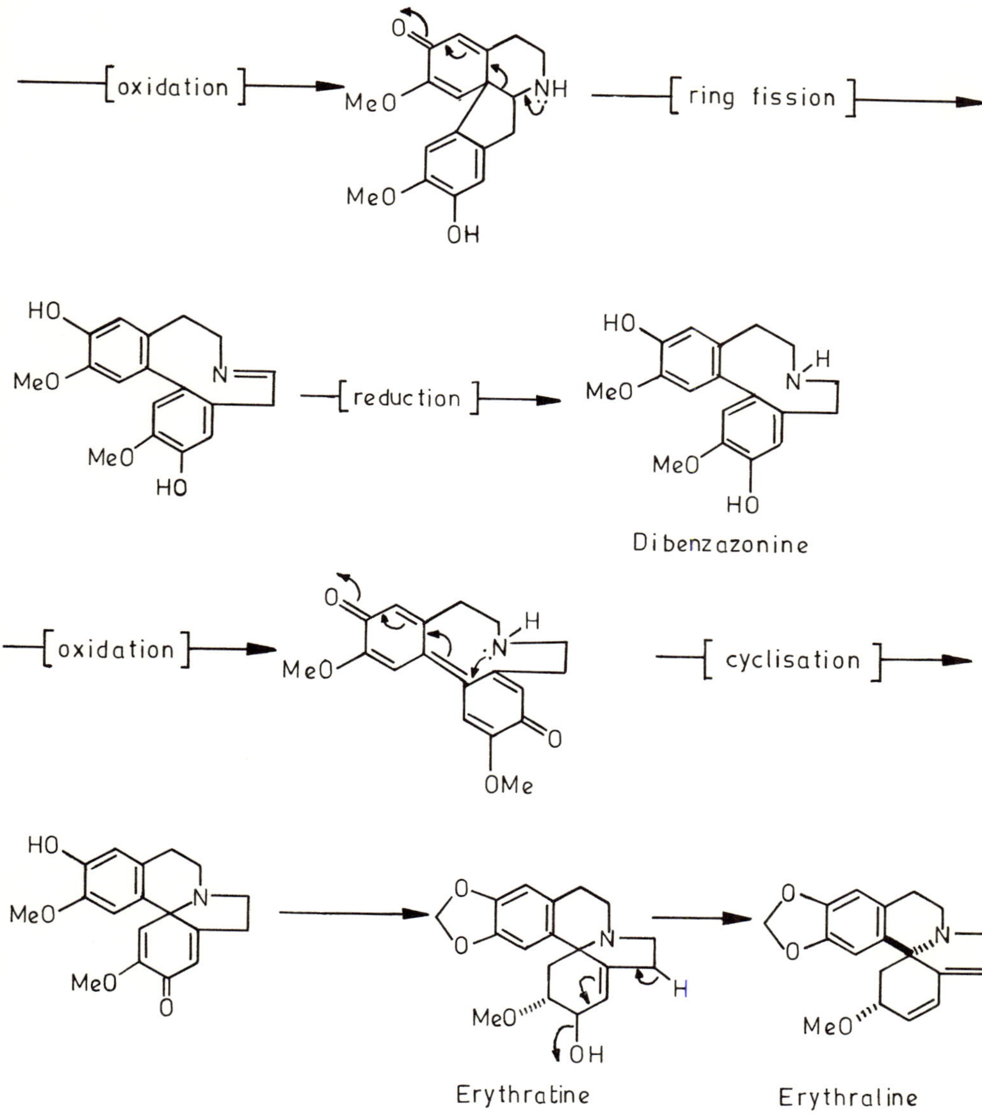
oxidation
MeO
NH
OH
ring fission
HO
MeO
N
HO
reduction
Dibenzazonine
oxidation
H
O
OMe
cyclisation
Erythratine
Erythraline

1. D. H. R. Barton, R. James, G. W. Kirby, D. W. Turner and D. A. Widdowson, Chem. Comm., 295 (1966); D. H. R. Barton, R. James, G. W. Kirby and D. A. Widdowson, *ibid*., 266 (1967); D.H.R. Barton, R. James, G. W. Kirby, D. W. Turner and D. A. Widdowson, J. Chem. Soc. (C), 1529 (1968); D. H. R. Barton, R. B. Boar and D. A. Widdowson, J. Chem. Soc. (C), 1213 (1970).

2. Only the (+)-isomer was utilized in erythraline formation in *Erythrina crista galli* plants.

GALANTHAMINE

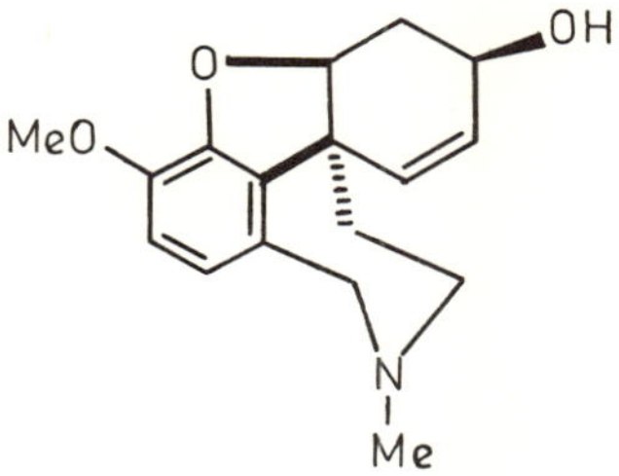

Galanthamine, one of the important members of the Amaryllidaceae family of alkaloids and possessing analgesic activity comparable to morphine[1] is biosynthesised[2] from tyrosine and phenylalanine following a pathway precisely as predicted by Barton[3] involving a crucial phenolic coupling step.

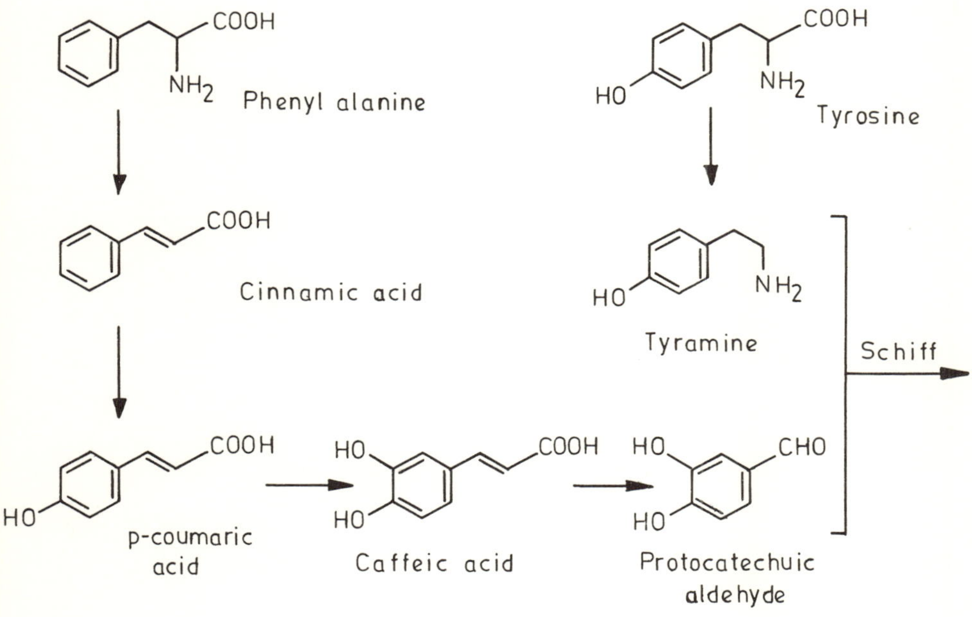

GALANTHAMINE

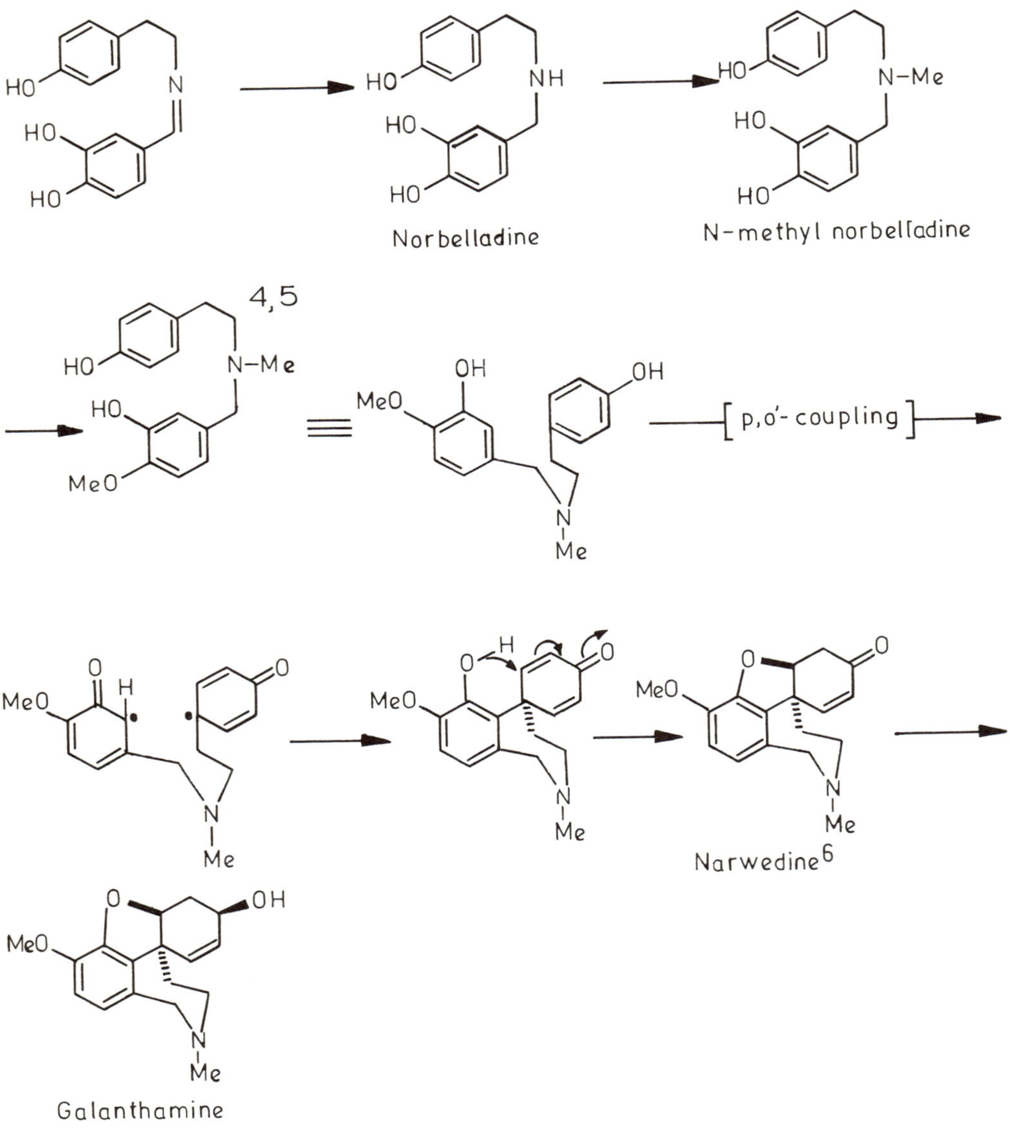

1. T. Kametani, C. Seino, K. Yamaki, S. Shibuya, K. Fukumoto, K. Kigasawa, F. Satoh, M. Huragi and T. Hayasaka, J. Chem. Soc., C, 1043 (1971).

2. D. H. R. Barton, G. W. Kirby, Proc. Chem. Soc., 392 (1960); J. Chem. Soc., 806 (1962); D. H. R. Barton, G. W. Kirby, J. B. Taylor, G. M. Thomas, Proc. Chem. Soc., 254 (1961); J. Chem. Soc., 4545 (1963).

3. D. H. R. Barton and T. Cohen in 'Festschrift Arthur Stoll' Birkhauser, Basle, 1957, p. 117.

4. The sequence norbelladine → N-methylnorbelladine → O,N-dimethylnorbelladine was established by the appropriate labelling experiments and the importance of this order of methylation was supported by the failure to observe any significant incorporation of O-methylnorbelladine into galanthamine in 'King Alfred' daffodils (Ref. 2).

5. O, N-dimethylnorbelladine on ferricyanide oxidation afforded narwedine which on reduction gave galanthamine (D. H. R. Barton and G. W. Kirby, J. Chem. Soc., 806 (1962)).

6. Labelled narwedine was incorporated into galanthamine (J. G. Bhandarkar and G. W. Kirby, J. Chem. Soc., C, 1224 (1970)).

GIBBERELLIC ACID

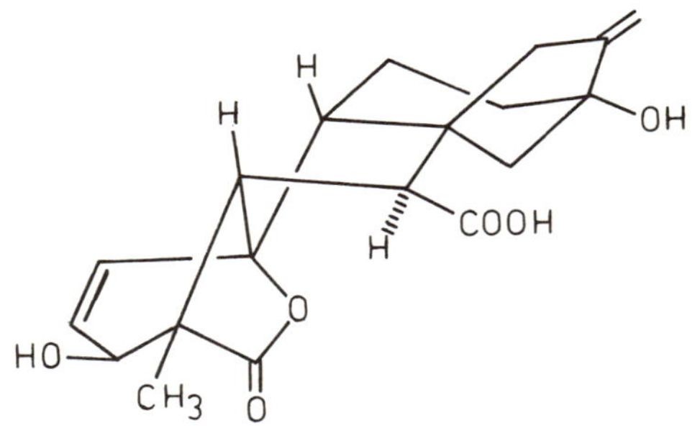

The biosynthesis[1] of the plant growth hormone, Gibberellic acid, is characterised by a plethora of fascinating molecular transformations that take place in a relatively rigid molecular framework.

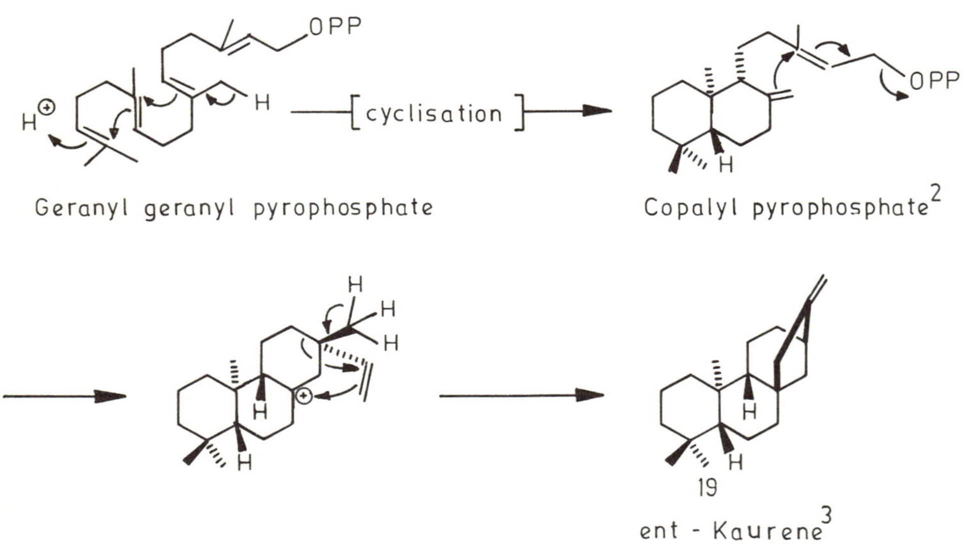

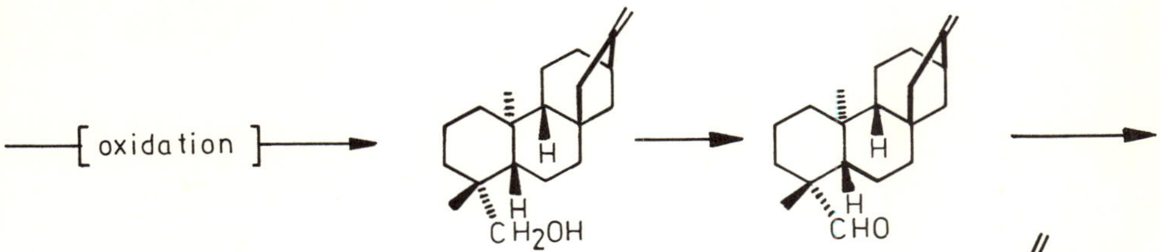

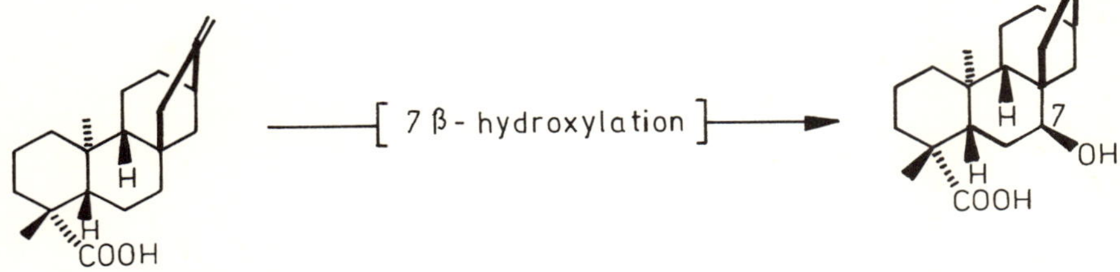

19

ent-Kaurenoic acid

7β-hydroxy-ent-Kaurenoic[4] acid

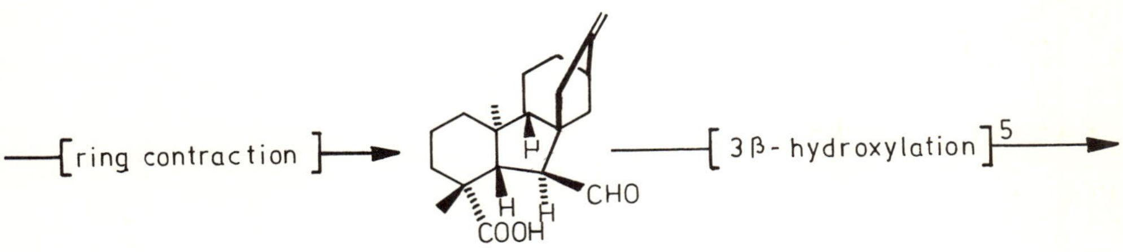

GA_{12}-aldehyde[6]

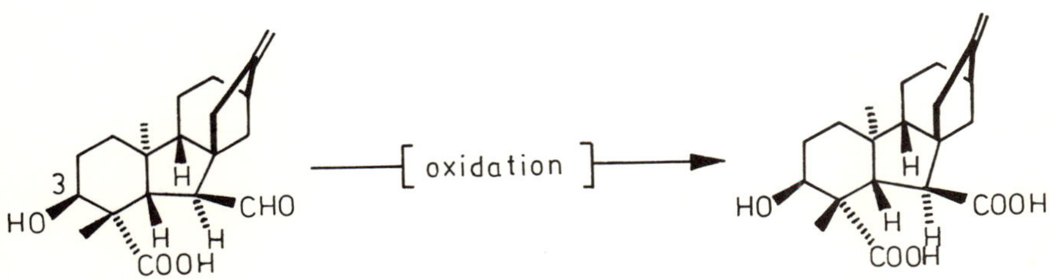

GA_{14}-aldehyde

GA_{14}

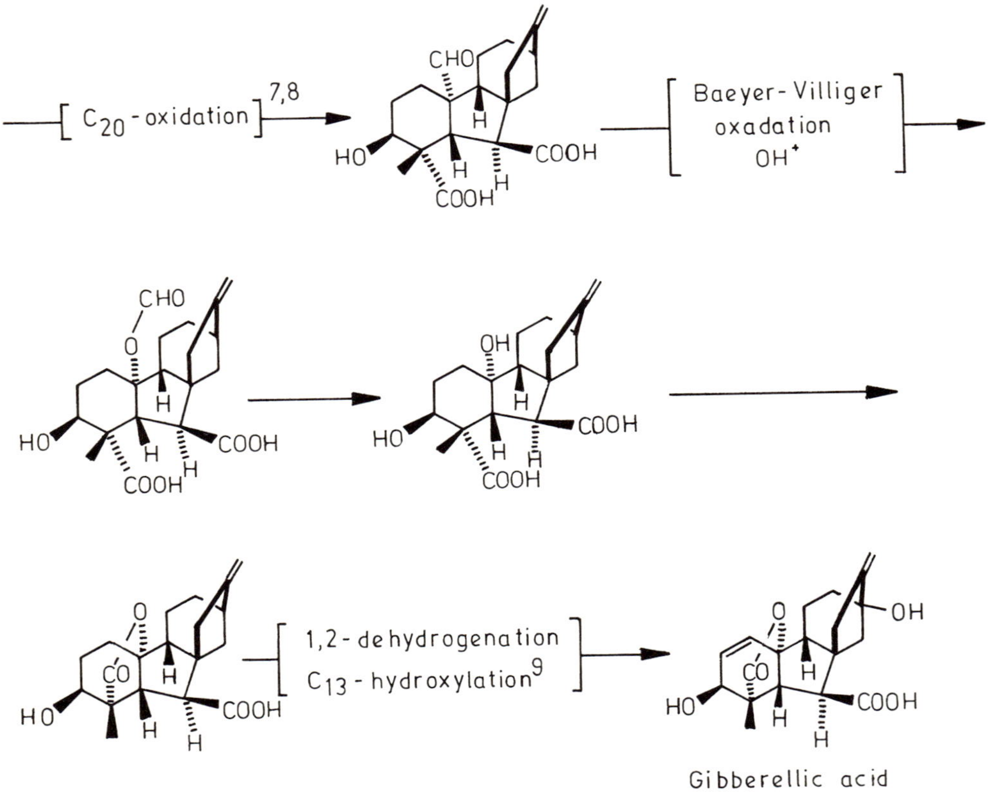

1. A. J. Birch, R. W. Richards, H. Smith, A. Harris and W. B. Whalley, Tetrahedron, 7, 241 (1959); B. E. Cross, in Progress in Phytochemistry, Vol. 1, eds. L. Rheinhold and Y. Liwschitz, John Wiley, New York, 1968, p. 195; A. Lang, Ann. Rev. Plant. Physiol., 21, 537 (1970); J. MacMillan, in 'Aspects of Terpenoid Chemistry and Biochemistry', ed. T. W. Goodwin, Academic Press, London and New York, 1971, p. 153; J. MacMillan and R. J. Pryce, in 'Phytochemistry', Vol. III,

ed. L. P. Miller, Van Nostrand Reinhold Co., New York, 1973 p. 283.

2. Copalyl pyrophosphate was shown to be a specific precursor of gibberellic acid in G. fujikuroi (J. R. Hanson and A. F. White, J. Chem. Soc. (C), 981 (1969)).

3. The stepwise oxidation of ent-kaurene to the 19-ol, 19-al, 19-oic acid and finally the 7β-hydroxy kaurenoic acid and their incorporation into Gibberellic acid has been demonstrated (Ref. 1). It is not possible to rationalise the oxidation of the unactivated 19-methyl group on the basis of in vitro experience. It is probable that the enzyme-coenzyme system responsible for this must be complexed with the olefinic function and that the oxidation is brought about by remote control (R. Breslow, Chem. Soc. Rev., 1, 553 (1972)). The conformational representation of ent-kaurene clearly indicates such possibilities.

4. It has been shown (J. R. Hanson and J. Hawker, Chem. Comm., 208 (1971)) that the 7β-hydroxy acid is formed from ent-kaurenoic acid by hydroxylation with retention of configuration at position 7. It was further suggested that its ring contraction to the ent-gibberellane aldehyde involves a hydride shift from C-6 to C-7 as the 7,8-bond migrates. The 6β-hydrogen is lost and the 6α-hydrogen is retained:

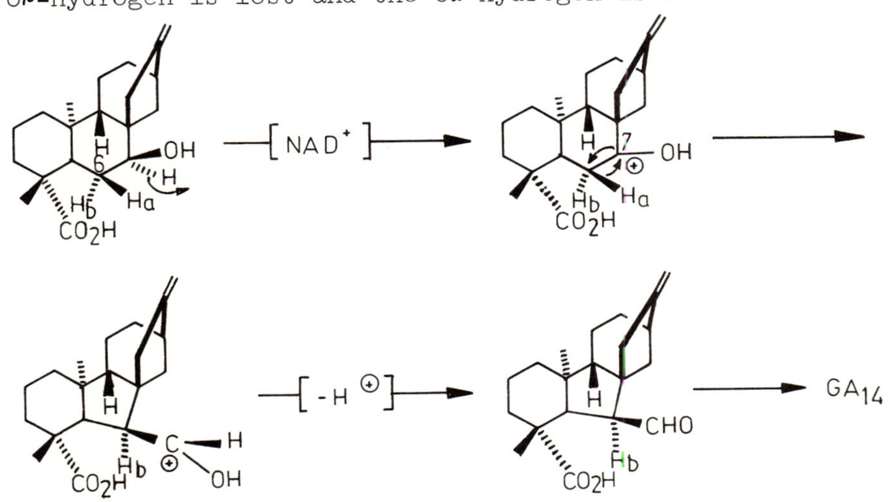

Recent experiments (J. E. Graebe, P. Hedden and J. MacMillan, Chem. Comm., 161 (1975)) with [6,6-$^3H_2$] - [17-^{14}C]ent-7α-hydroxykaurenoic acid have confirmed the above proposition.

5. The hydroxylation of carbon centres located β to a functional group is quite common in biosynthesis (e.g. colchicine, haemanthamine) and perhaps takes place <u>via</u> sequence, fragmentation, oxidation and recombination:

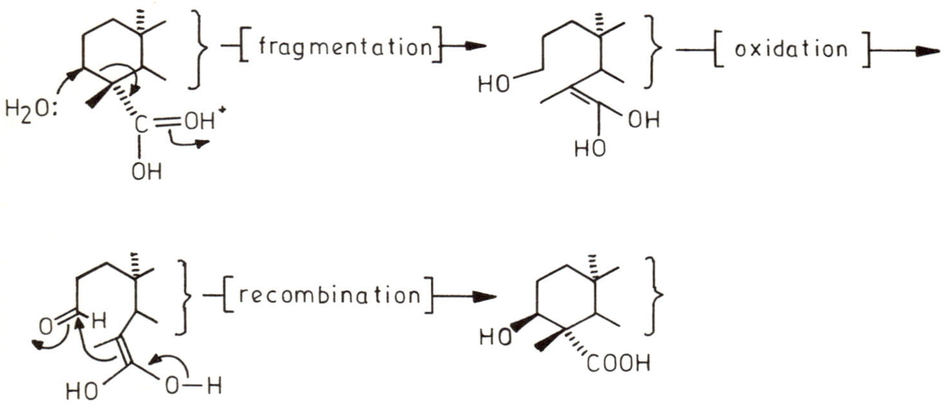

6. The sequence GA_{12}-aldehyde → GA_{14}-aldehyde → GA_{14} has been established in cultures of *G. fujikuroi*. (P. Hedden, J. MacMillan and B. O. Phinney, J. Chem. Soc., Perkin I, 587 (1974).)

7. It is interesting to observe that carbon centres 7 and 20 which are precisely four carbons away, are oxidised. Only in these two locations, the carboxyl function can reach the methyl hydrogens. It is quite possible that these oxidations arise from a peracid intermediate:

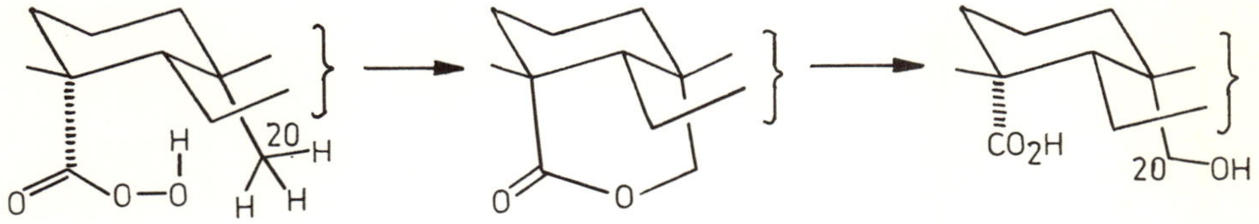

8. The loss of C-20 may occur at the aldehyde level by the biochemical equivalent of a Baeyer-Villiger oxidation followed by hydrolysis of the resulting ester as suggested by Hanson (Ref. 2).

9. The bridgehead hydroxylation can be readily rationalised:

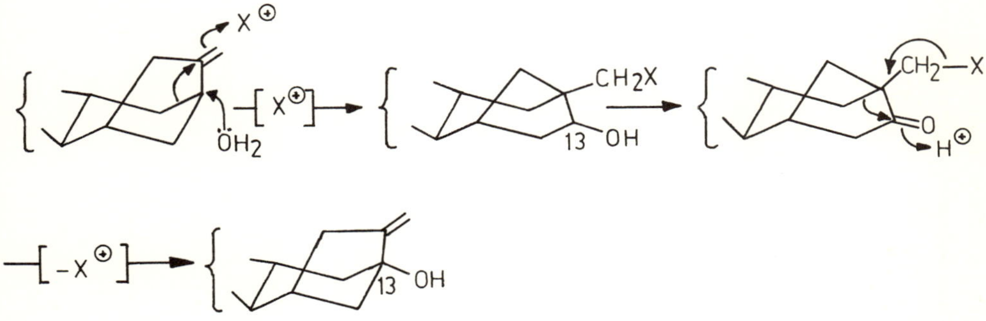

GRISEOFULVIN

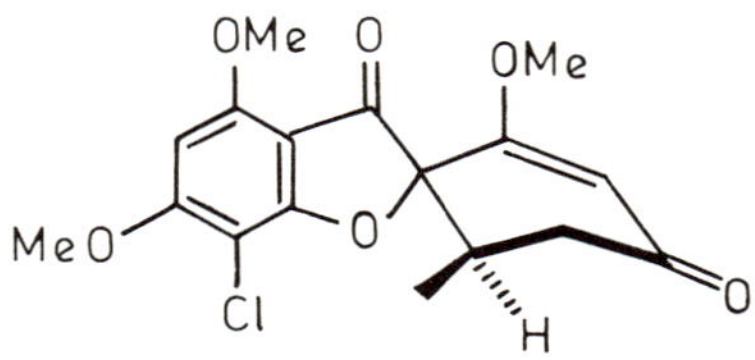

The unusual carbon framework of the potent antifungal agent Griseofulvin is generated from acetyl-CoA <u>via</u> sequence polyketide cyclisation, halogenation and phenolic coupling.[1,2]

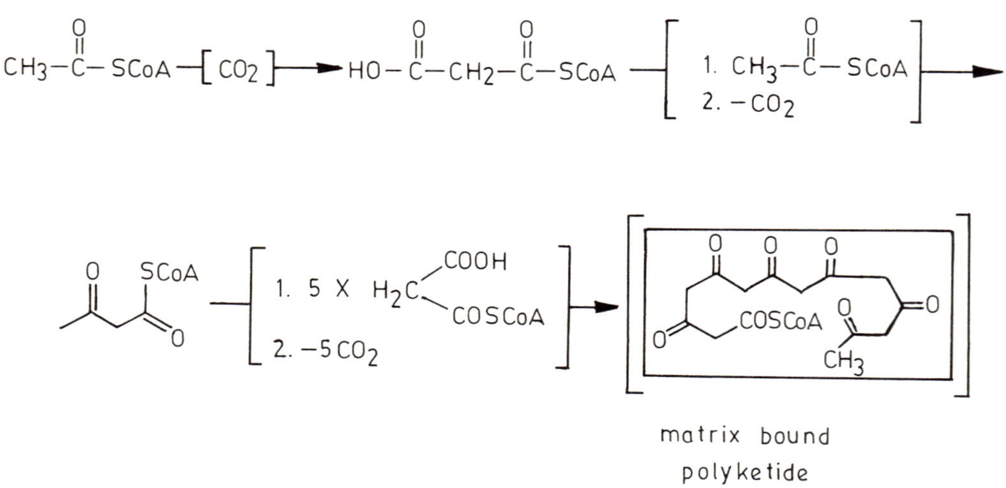

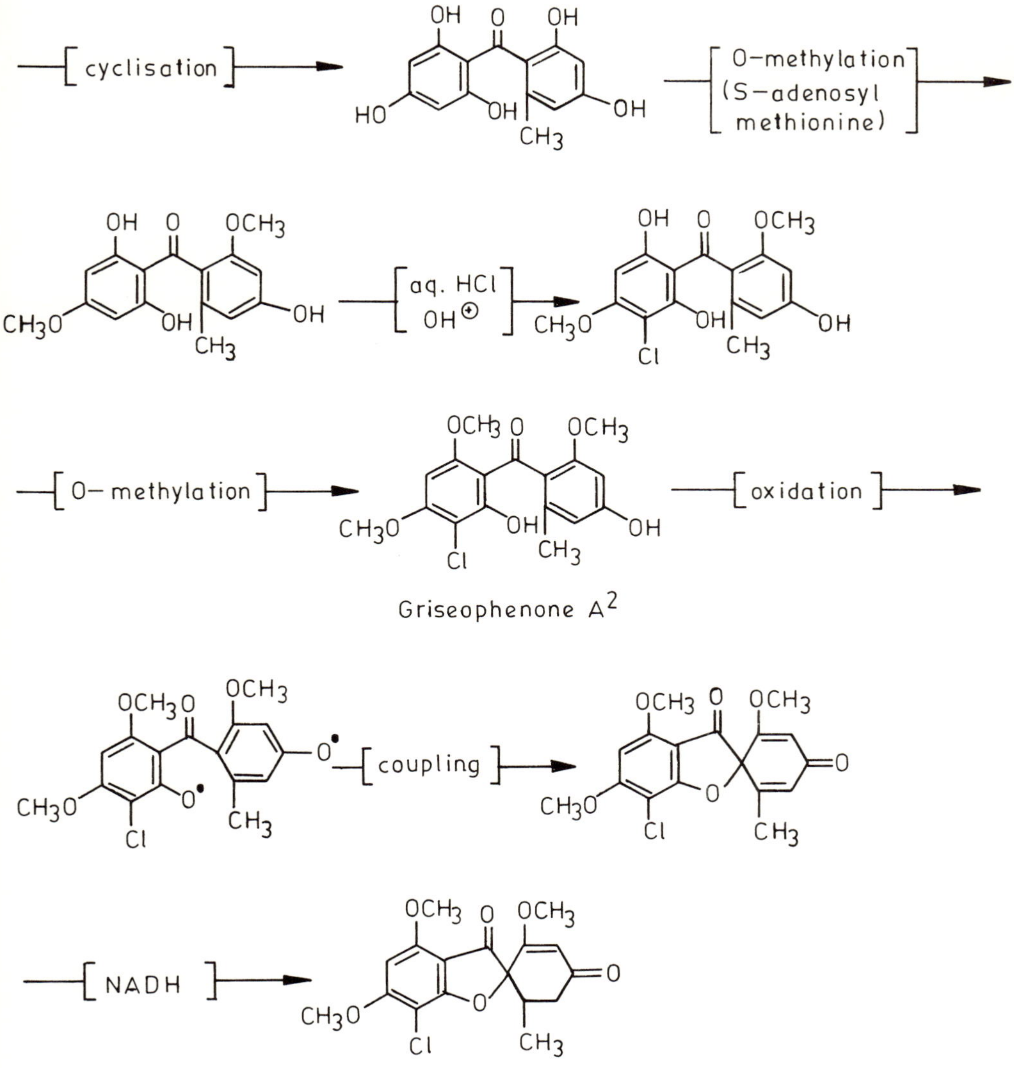

Griseofulvin

1. A. I. Scott in 'Oxidative Coupling of Phenols', eds. W. I. Taylor and A. R. Battersby, Marcel Dekker, Inc., New York, 1967, p. 95; J. F. Grove in 'Antibiotics, Vol. II', eds. D. Gottlieb and P. D. Shaw, Springer Verlag, Berlin, 1967, p. 123.

2. The in vitro coupling of Griseophenone A to dehydrogriseofulvin with ferricyanide has been demonstrated (A. C. Day, J. Nabney and A. I. Scott, J. Chem. Soc., 4067 (1961); D. Taub, C. H. Kuo, H. L. Slates and N. L. Wendler, Tetrahedron, 19, 1 (1963)).

HAEMANTHAMINE

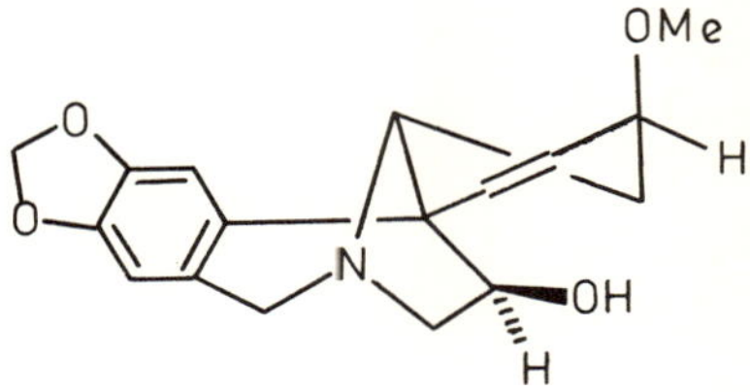

Haemanthamine is biosynthesised[1] in Amaryllidaceae plants from tyrosine and phenylalanine via the intermediacy of norbelladine and O-methylnorbelladine by a p,p' phenolic coupling.[2] The initially formed coupled product is transformed to the Oxocrinine skeleton by a Michael addition. The Oxocrinine ⟶ Haemanthamine change involves a fascinating hydroxylation as is found to take place in the biosynthesis of Colchicine.

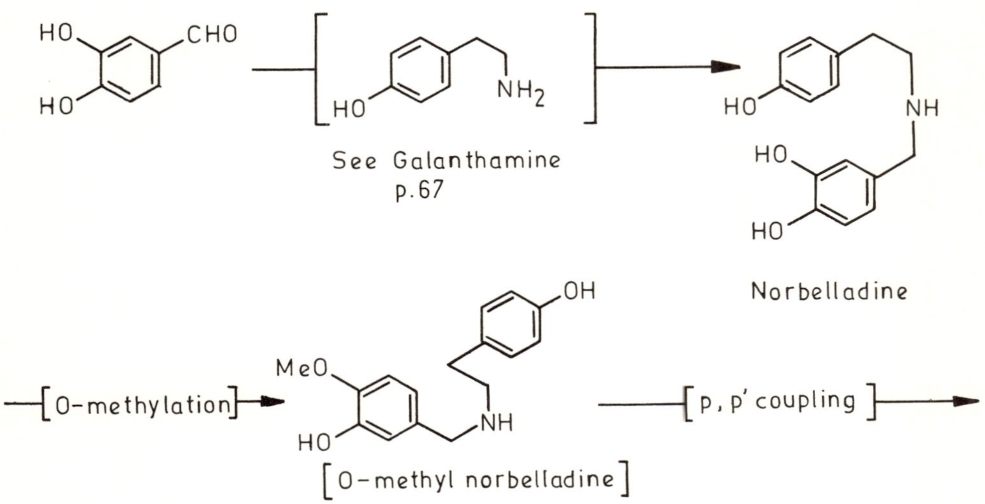

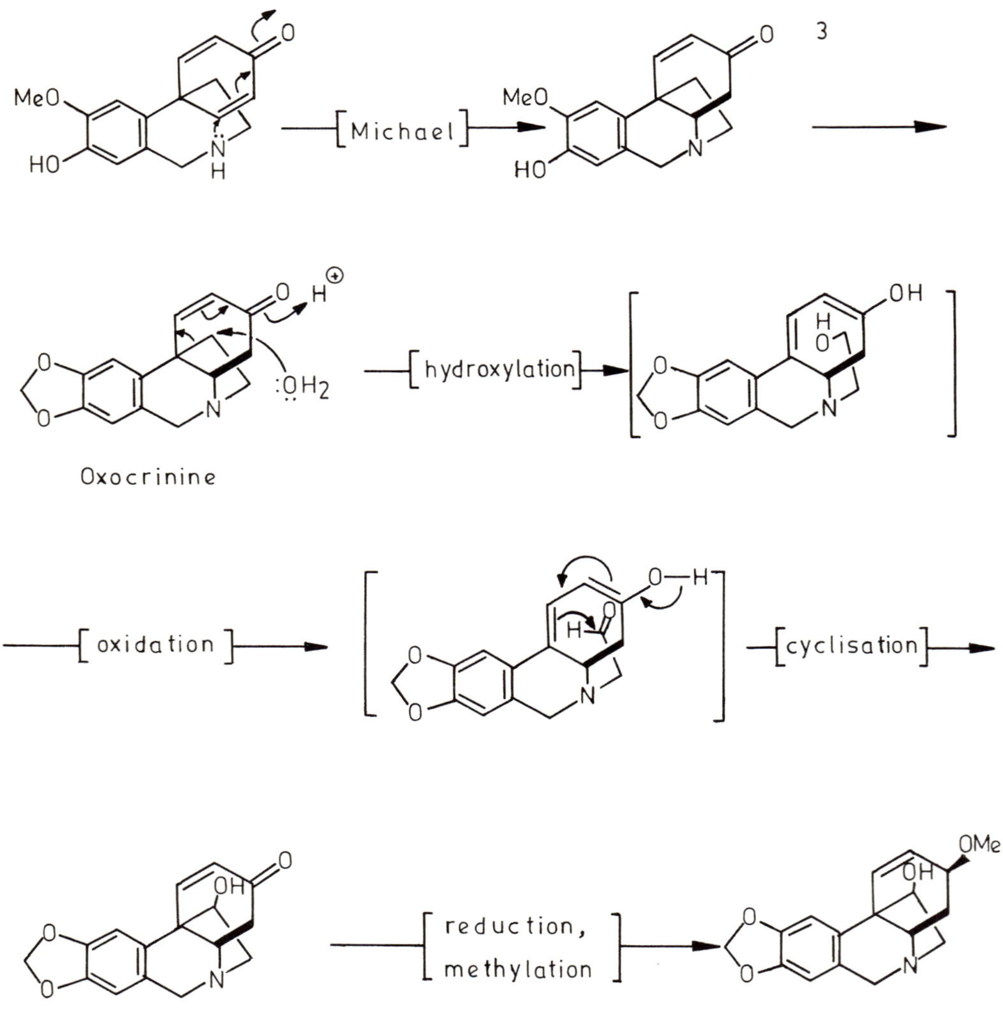

Haemanthamine

1. W. C. Wildman, H. M. Fales, R. J. Highet, S. W. Breuer and A. R. Battersby, Proc. Chem. Soc., 180 (1962); D. H. R. Barton, J. W. Kirby, J. B. Taylor and G. M. Thomas, J. Chem. Soc. 4545 (1963); A. R. Battersby, R. Binks, S. W. Breuer, H. M. Fales, W. C. Wildman and R. J. Highet, J. Chem. Soc., 1595 (1964); A. R. Battersby in 'Oxidative Coupling of Phenols', Marcel Dekker, Inc., New York, 1967, ed. W. I. Taylor and A. R. Battersby, p. 119.

2. The biosynthetic experiments were designed based on the important proposal of Barton (D. H. R. Barton and T. Cohen in 'Festschrift Arthur Stoll', Birkhauser, Basle, 1957, p. 117) that a key step in the biosynthesis of Amaryllidaceae alkaloids is the oxidative coupling of intermediates such as norbelladine.

3. Using doubly labelled O-methylnorbelladine [$O-{}^{14}CH_3$];[$1-{}^{14}C$], it has been proved that methylenedioxy group of haemanthamine is formed in nature by cyclisation of the O-methoxyphenol group. Such a ring closure could occur via the phenoxy radical or the corresponding cation as shown below.

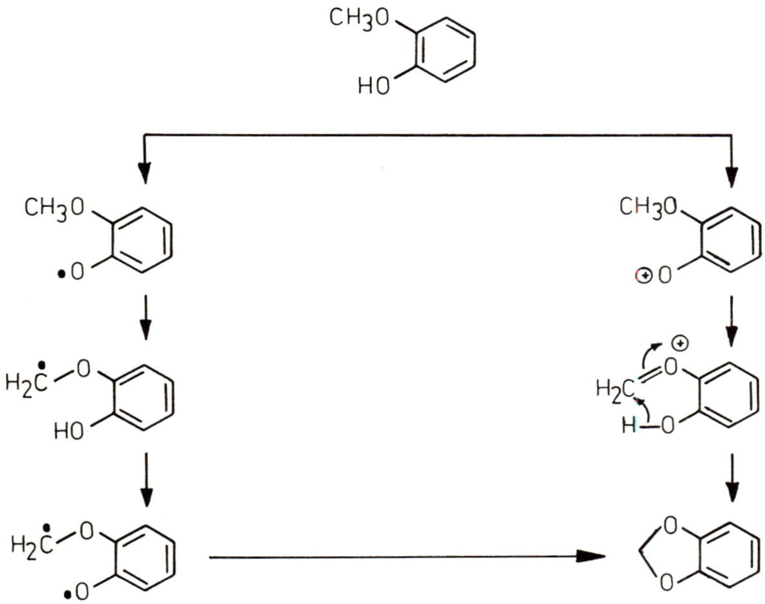

HISTIDINE

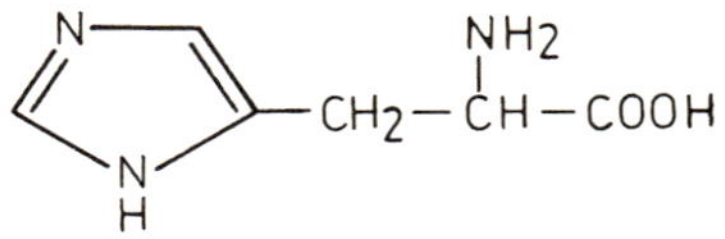

The notion that apparently complicated transformations take place only under the artificial conditions available in the laboratory is far from correct. Indeed, in Nature such processes are common and their genesis could be related to the characteristic feature of the art in biosynthesis, namely, the maximum optimisation in the construction of molecular frameworks, particularly those which play a vital role in life processes. This facet is best illustrated with 'ATP-Imidazole' cycle, which is related not only to the biosynthesis of histidine, but also to guanine triphosphate (GTP) and adenosine triphosphate (ATP).

The heart of Histidine is the imidazole system. Nature makes the imidazole unit of Histidine using another imidazole as a template and via a cyclic operation that regenerates the template imidazole:[1]

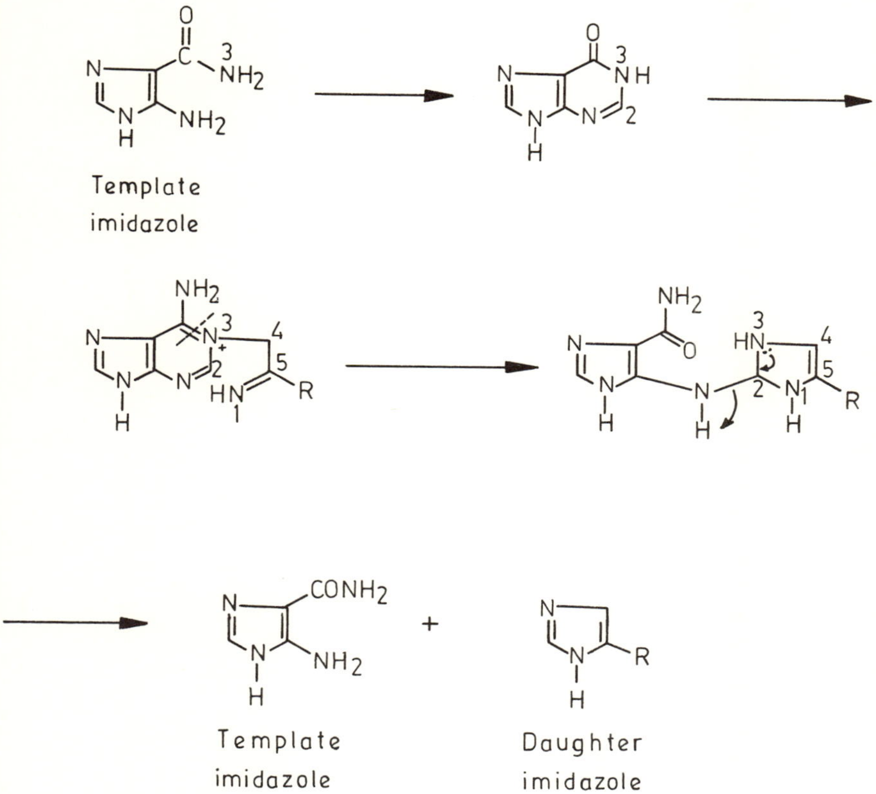

The formation of the 'daughter' imidazole can be best represented by the ATP-imidazole cycle:

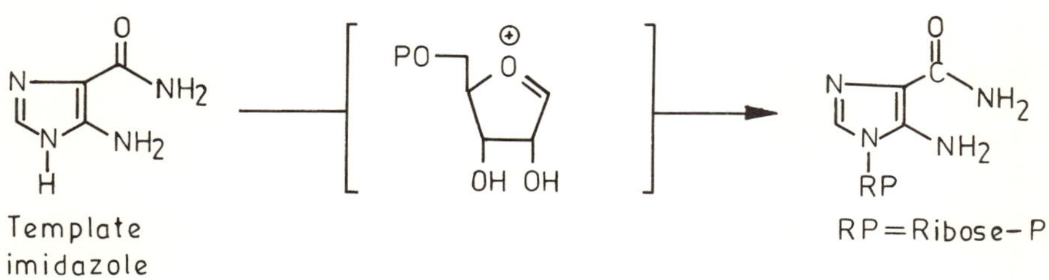

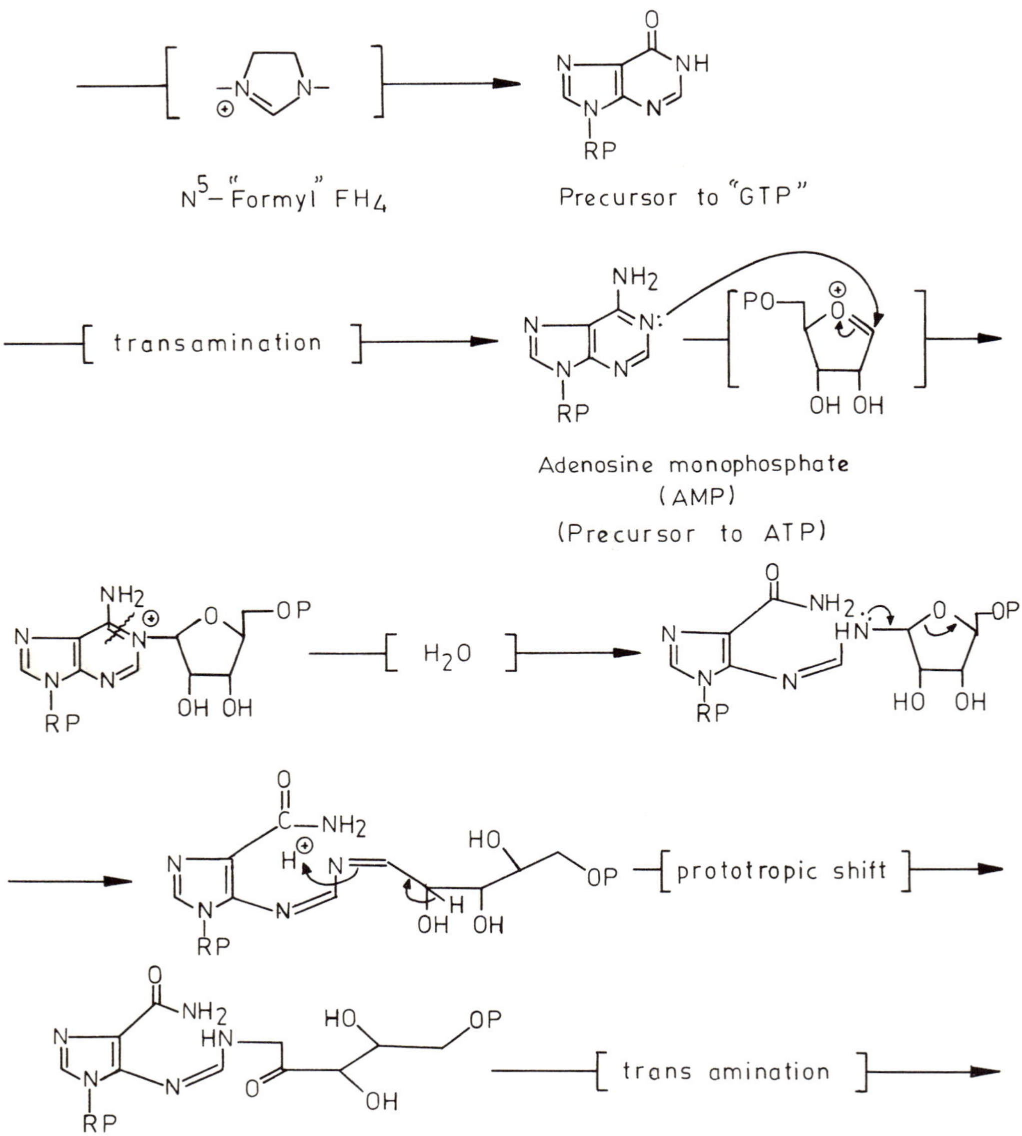

N5–"Formyl" FH4
Precursor to "GTP"
transamination
Adenosine monophosphate
(AMP)
(Precursor to ATP)
H2O
prototropic shift
trans amination

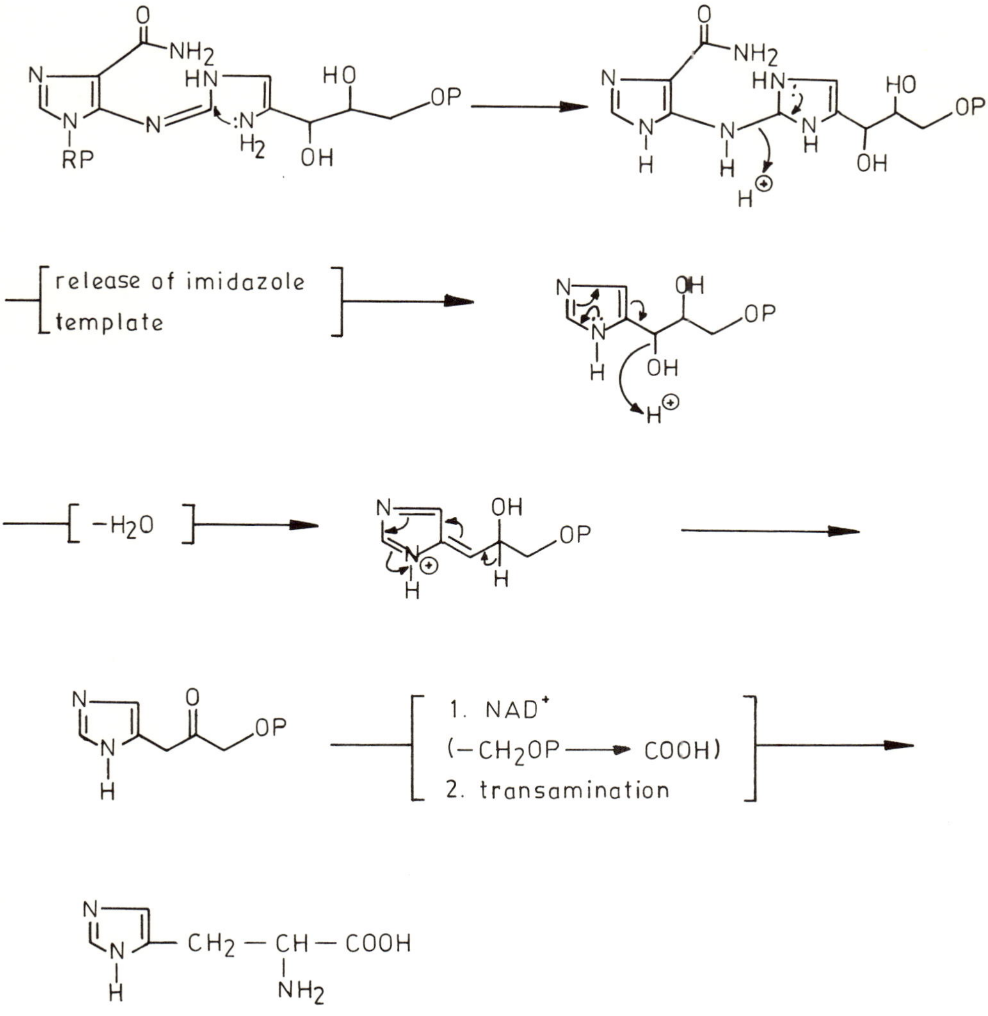

Histidine

1. P. Bernfeld, Biogenesis of Natural Products, Pergamon, Oxford, 1963, p.21.

ILLUDIN-M

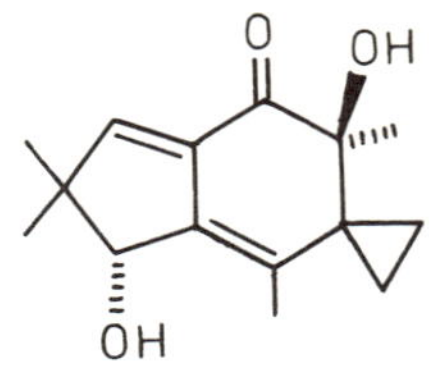

Certain types of poisonous, bioluminescent mushrooms yield the antitumor substance, Illudin. Nature creates the unusual illudin skeleton from humulene by a fascinating and breathtaking array of molecular rearrangements![1]

2

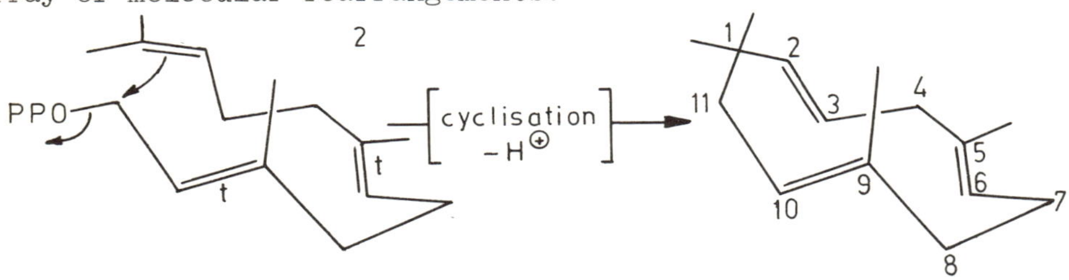

EE-Farnesyl pyrophosphate

Humulene

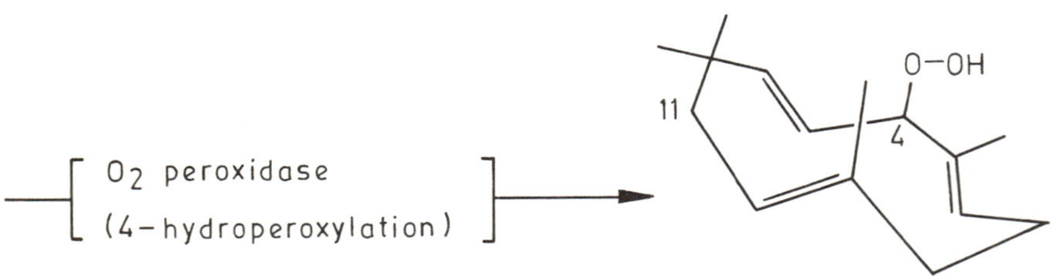

3

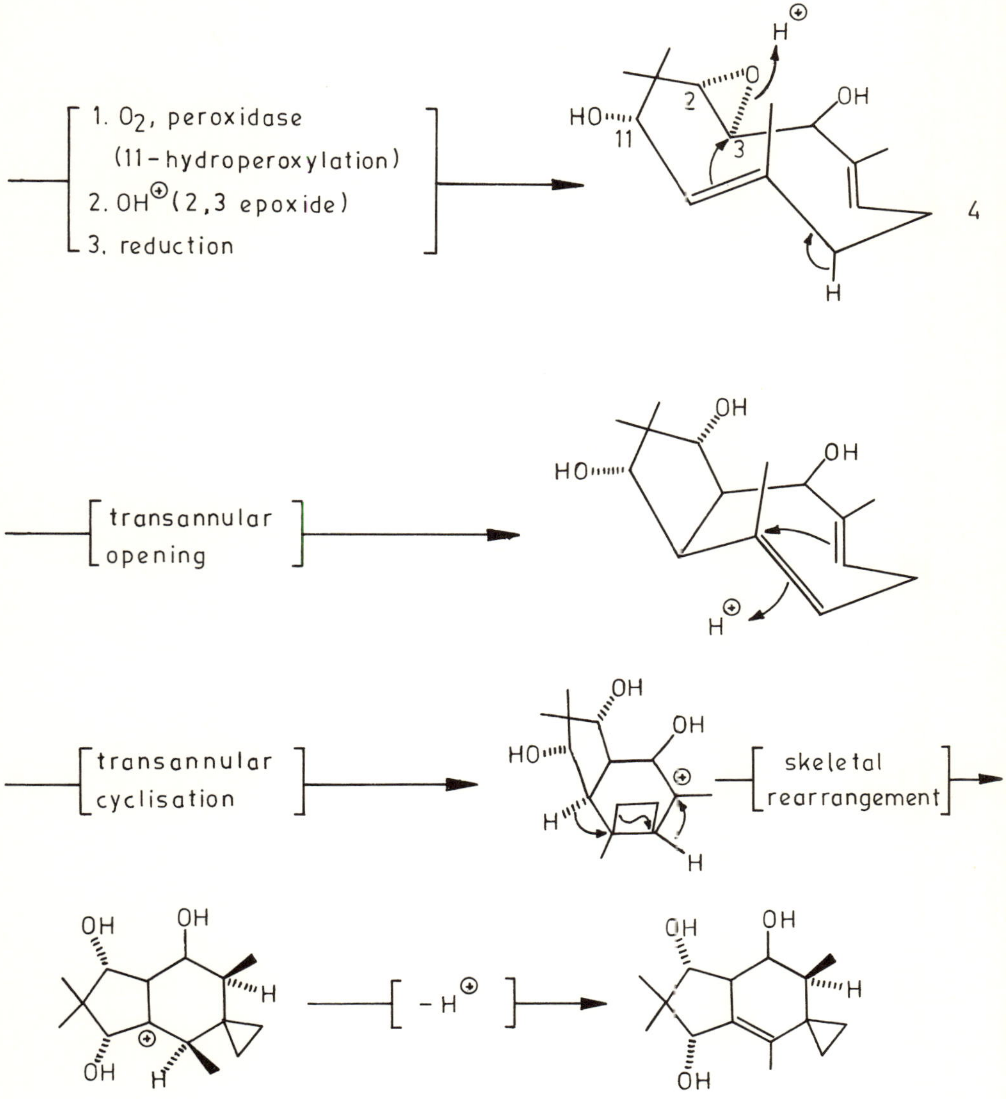

1. O_2, peroxidase
(11-hydroperoxylation)
2. $OH^{\oplus}$ (2,3 epoxide)
3. reduction
HO
11
2
3
O
$H^{\oplus}$
OH
H
4
transannular
opening
OH
HO
OH
$H^{\oplus}$
transannular
cyclisation
OH
HO
OH
H
H
skeletal
rearrangement
OH
OH
H
OH
H
$-H^{\oplus}$
OH
OH
H
OH

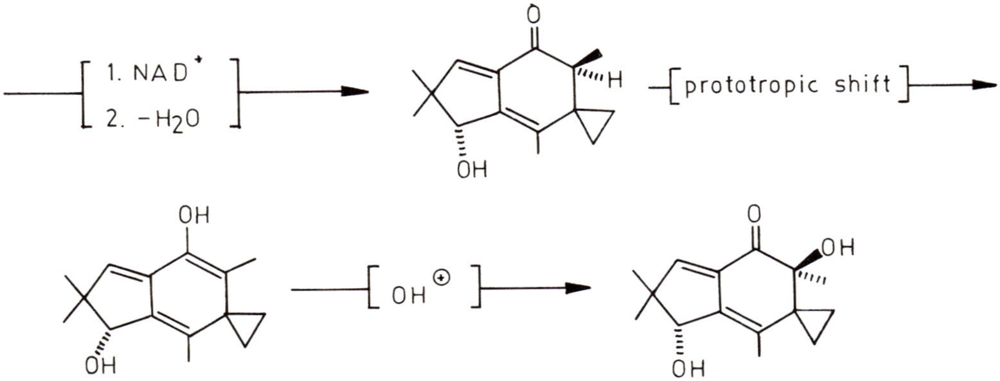

Illudin M

1. T. McMorris and M. Anchel, J. Amer. Chem. Soc., 87, 1594 (1965); K. Nakanishi, M. Ohashi, M. Tada and Y. Yamada, Tetrahedron, 21, 1231 (1965); M. Anchel, T. McMorris and M. Singh, Phytochemistry, 9, 2339 (1970); J. R. Hanson and T. Marten, Chem. Comm., 171 (1973).

2. [2-$^{3}H_2$],[2-^{14}C] doubly labelled mevalonate was successfully incorporated into Illudin-M in Clitocybe illudens.

3. Dehydration of this intermediate leads to Zerumbone.

4. Three -[2-$^{3}H_2$] and one [4R-4-^{3}H] mevalonoid hydrogen atoms are incorporated into illudin M. The loss of a [2-$^{3}H_2$] mevalonoid label in the cyclopropane ring is in accord with the non-concerted cyclisation of humulene to form illudin.

INSULIN

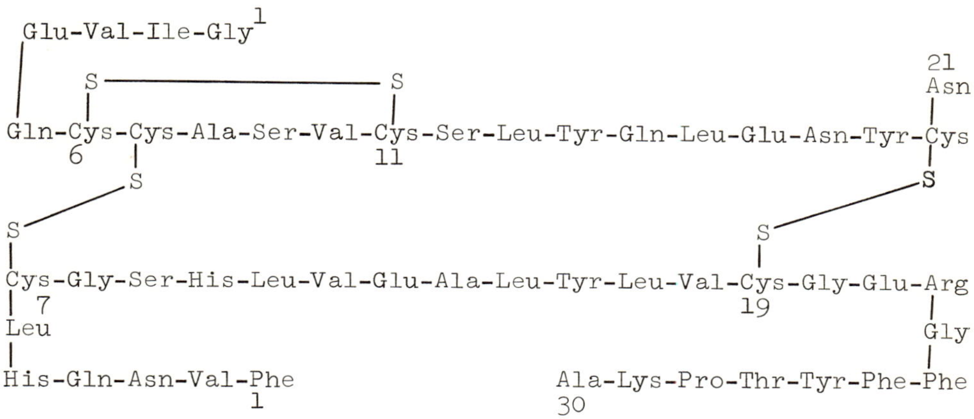

The biosynthesis[1] of Insulin, unravelled much after it was synthesised[2] by man, reflects clearly the superiority of Nature in the creation of molecular frameworks. The intramolecular processes of forming the delicate disulfide bridges in proinsulin[3], the natural precursor of insulin, are to be contrasted with all the <u>in vitro</u> preparations of this hormone involving the relatively inefficient molecular combinations of the two polypeptide chains.

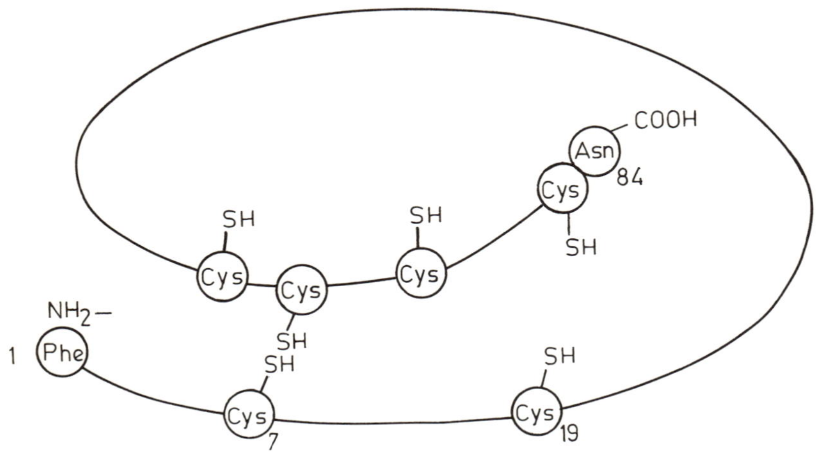

Nascent protein

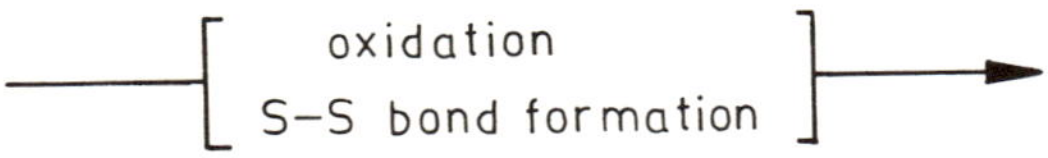

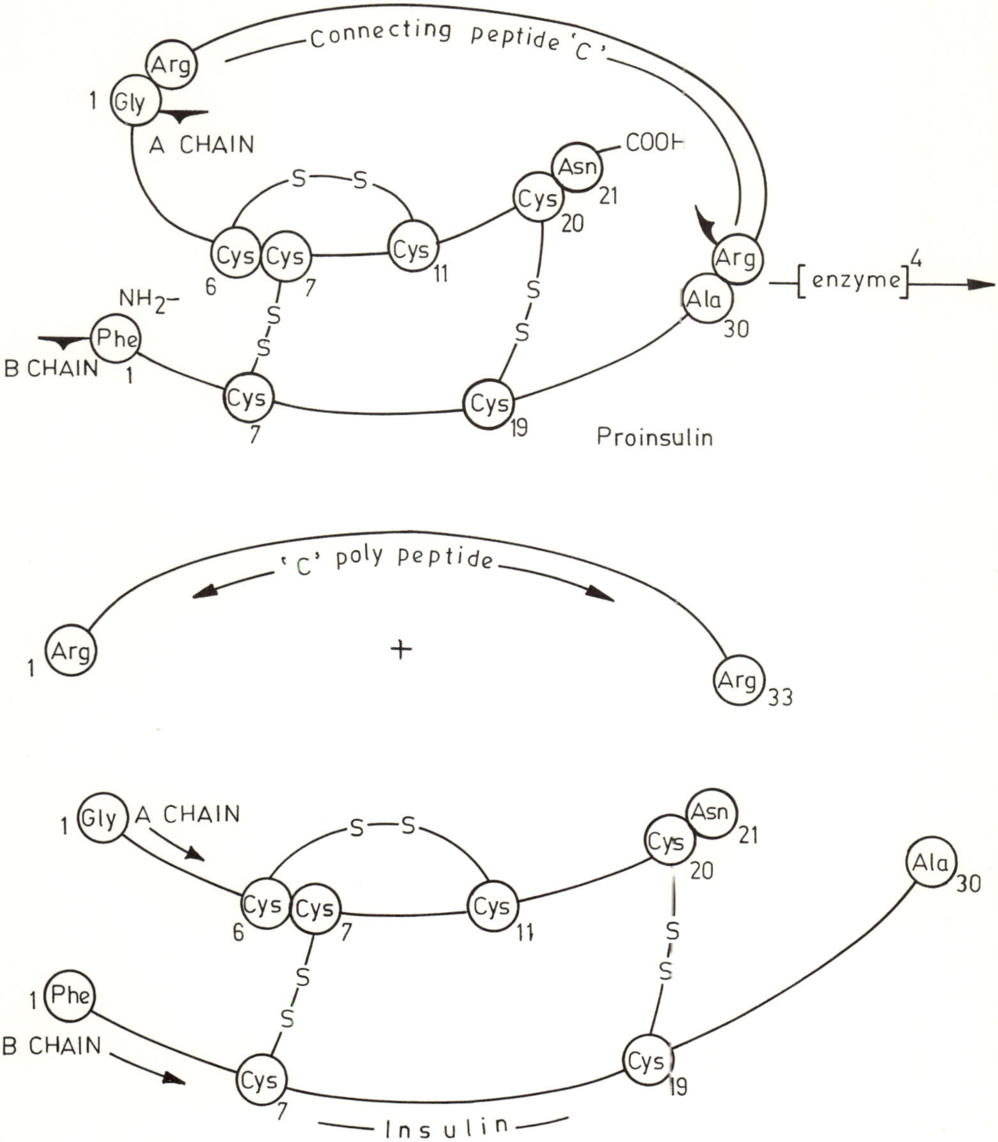
Connecting peptide 'C'
Arg
1 Gly
A CHAIN
COOH
Asn 21
Cys 20
S—S
Cys 6
Cys 7
Cys 11
S
S
Arg 4
Ala 30
[enzyme]
NH2–
Phe 1
B CHAIN
Cys 7
Cys 19
Proinsulin
'C' poly peptide
1 Arg
+
Arg 33
1 Gly A CHAIN
S—S
Asn 21
Cys 20
Ala 30
Cys 6
Cys 7
Cys 11
1 Phe
B CHAIN
Cys 7
Cys 19
Insulin

1. D. F. Steiner and P. E. Oyer, Proc. Nat. Acad. Sci., U.S.A., 57, 473 (1967); D. F. Steiner, D. D. Cunningham, L. Spigelman and B. Aten, Science, 157, 697 (1967); P. T. Grant and K. B. M. Reid, Biochem. J., 110, 298 (1968); A. K. Tung and C. C. Yip, Proc. Nat. Acad. Sci., U.S.A., 63, 442 (1969); A. C. Trakatellis and G. P. Schwartz, Nature, 225, 548 (1970); R. E. Chance, R. M. Ellis and W. W. Bromer, Science, 161, 165 (1968); D. F. Steiner, S. Cho, P. E. Oyer, S. Terris, J. D. Peterson and A. H. Rubinstein, J. Biol. Chem., 246, 1365 (1971).

2. A. Marglin and R. B. Merrifield, J. Amer. Chem. Soc., 88, 5051 (1966).

3. Proinsulin contains the A and B chains of insulin and a loop of 33 amino acid residues linking these chains. The Amino acid sequence of this loop of connecting peptide was found to be: Arg-Arg-Gly-Ala-Gln-Asn-Pro-Gln-Ala-Gly-Ala-Val-Glu-Leu-Gly-Gly-Gly-Leu-Gly-Gly-Leu-Gln-Ala-Leu-Ala-Leu-Glu-Gly-Pro-Pro-Gln-Lys-Arg.

4. The enzyme that catalyses the conversion of bovine proinsulin to intact insulin has been isolated and purified; the connecting peptide 'c' was released intact (C. C. Yip, Proc. Nat. Acad. Sci., U.S.A., 68, 1312 (1971).

LOGANIN, SECOLOGANIN

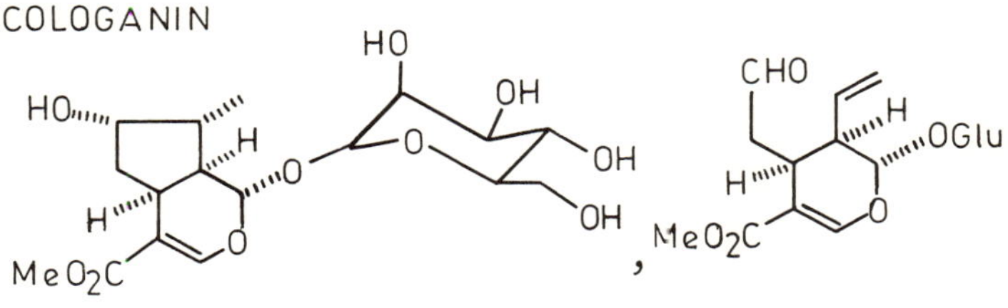

Loganin   Secologanin

Loganin, the molecule which plays a central role in the biosynthesis of indole alkaloids[1] is created in Nature from Geraniol.[2] Secologanin, the immediate precursor of the non-tryptophan part of indole alkaloids, is obtained from Loganin by an interesting 'retro-Prins' reaction.[3]

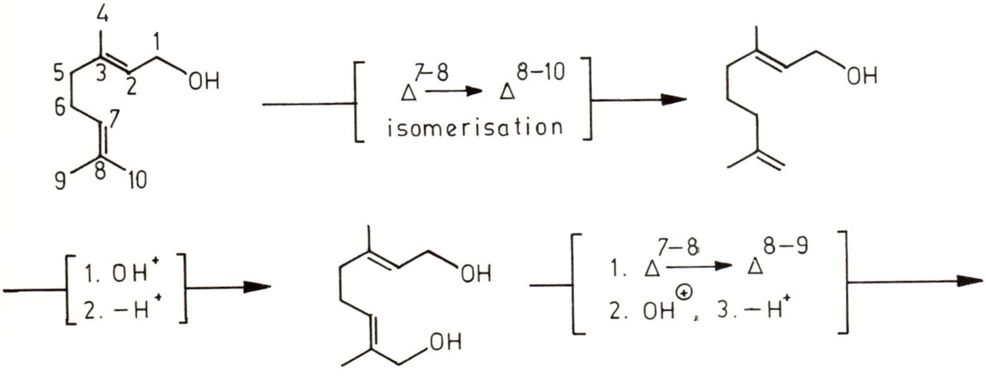

10-hydroxygeraniol[4]

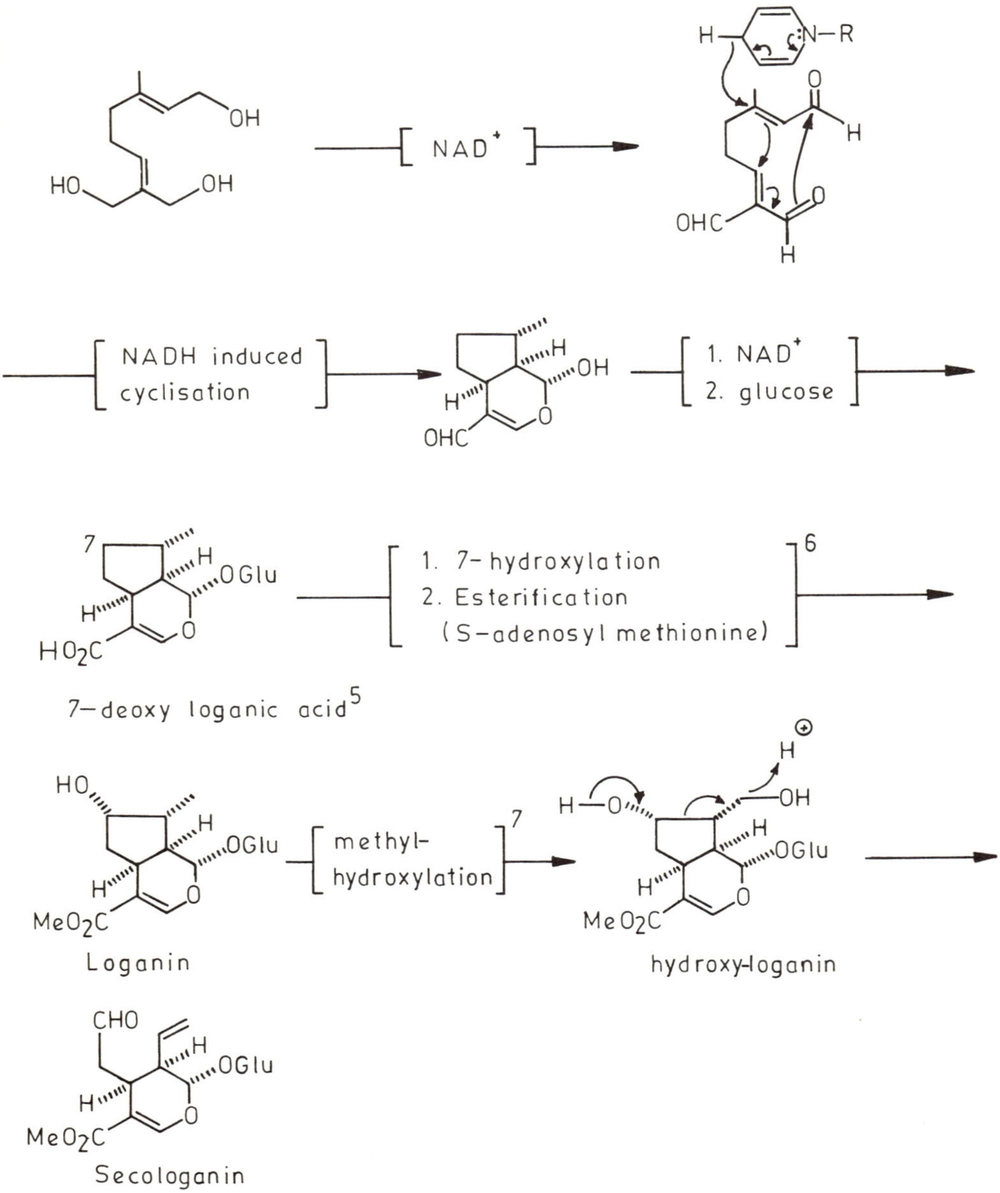
NAD+
NADH induced cyclisation
1. NAD+
2. glucose
1. 7-hydroxylation
2. Esterification
(S-adenosyl methionine)
6
7-deoxy loganic acid5
methyl-hydroxylation
7
Loganin
hydroxy-loganin
Secologanin

1. A. R. Battersby, Pure and Appl. Chem., 14, 117 (1967); D. V. Banthorpe, B. V. Charlwood and M. J. O. Francis, Chem. Rev., 72, 135 (1972).

2. A. R. Battersby, R. S. Kapil and R. Southgate, Chem. Comm., 131 (1968); S. Brechbuhler-Bader, C. J. Coscia, P. Loew, Ch. von Szczepanski and D. Arigoni, Chem. Comm., 136 (1968); C. J. Coscia and R. Guarnasccia, Chem. Comm., 138 (1968).

3. A. R. Battersby, A. R. Burnett and P. G. Parsons, J. Chem. Soc. (C), 1187 (1969).

4. [9-^{14}C] and [1-$^{3}H_2$]-10-hydroxy derivatives of geraniol and nerol were both incorporated into loganin in *V. rosea* plants. This is not surprising in view of the ready and reversible geraniol ⟶ nerol change:

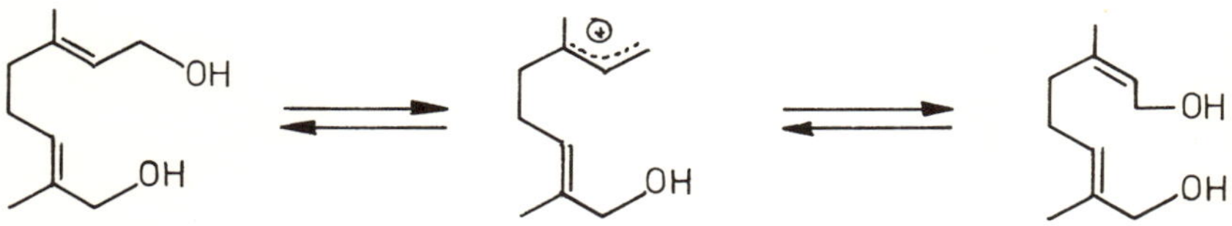

10-hydroxy-geraniol

10-hydroxy nerol

(E. Escher, P. Loew and D. Arigoni, Chem. Comm., 823 (1970); A. R. Battersby, S. H. Brown and T. G. Payne, Chem. Comm., 827 (1970).)

5. 7-deoxy loganic acid has been shown to be the bio-intermediate to loganin in *S. caroliniensis* plants (C. J. Coscia and R. Guarnaccia, Chem. Comm., 138 (1968)).

6. This unexpected α-hydroxylation can be readily rationalised on the basis of a conformationally favourable 1,5-hydride shift:

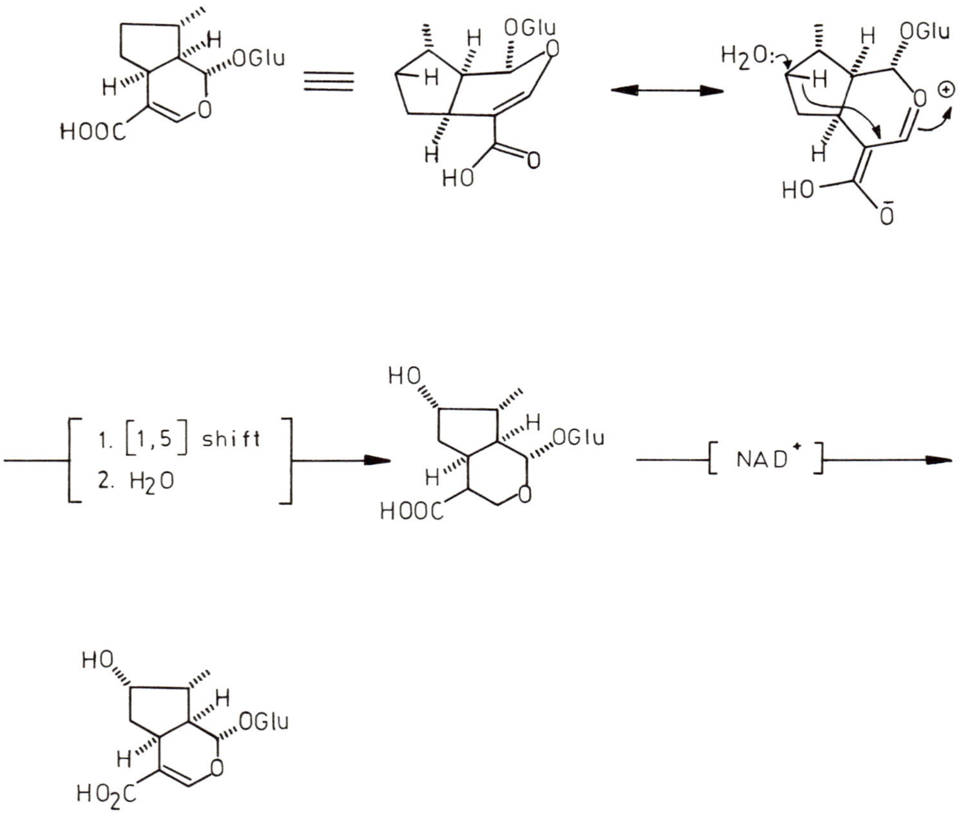

7. This hydroxylation of an 'unactivated methyl' perhaps takes place via an oxidation, hydration and reduction sequence:

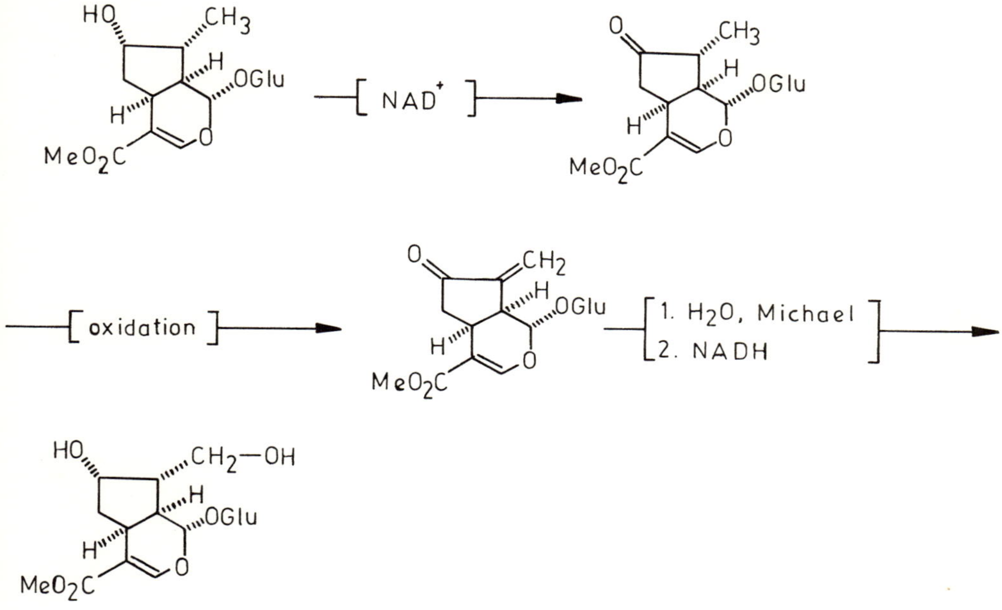

hydroxy loganin

LONGIFOLENE

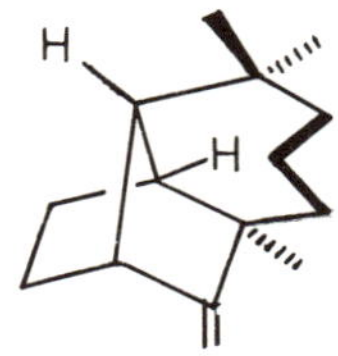

The biosynthesis of Longifolene[1,2] is illustrative of the deep seated carbonium ion rearrangements that often are associated with the construction of carbon frameworks, particularly in the area of isoprenoids.

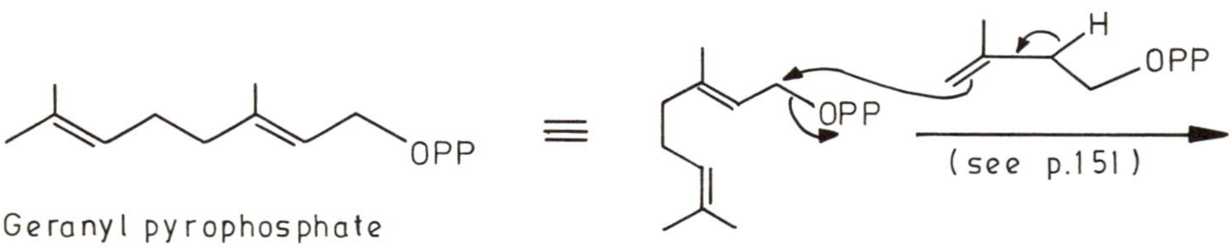

Geranyl pyrophosphate
(see p.151)

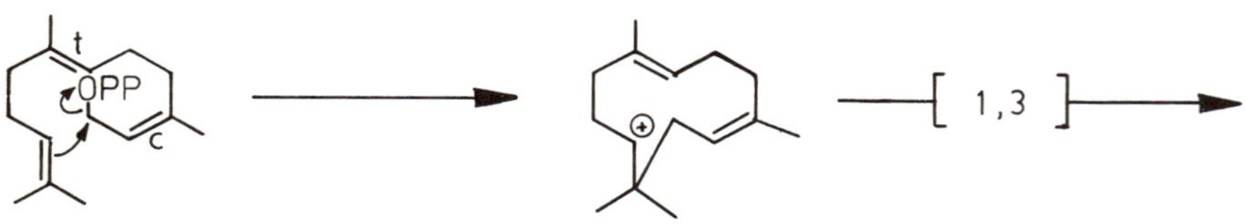

E,Z - Farnesyl pyrophosphate

LONGIFOLENE

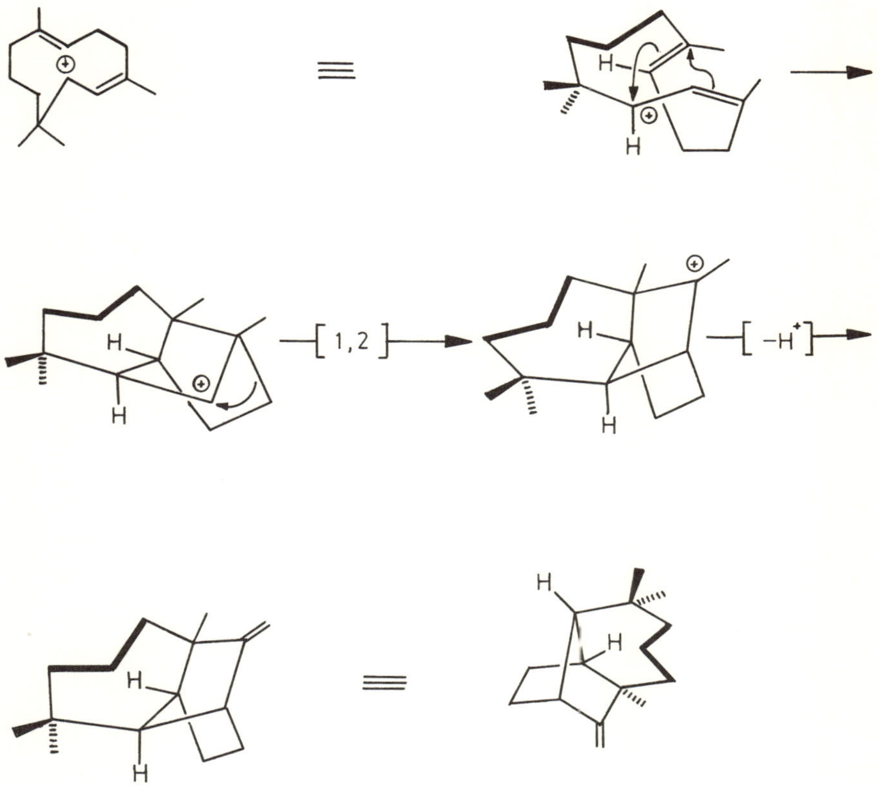

1. W. Parker, J. S. Roberts and R. Ramage, Quart. Rev., 21, 331 (1967) and references cited therein.

2. Labelling experiments with [1-^{14}C] acetate supported the above hypothesis in P. longifolia Roxb. (W. Sandermann and K. Bruns, Tetrahedron Letters, 261 (1962)).

LYCOPODINE

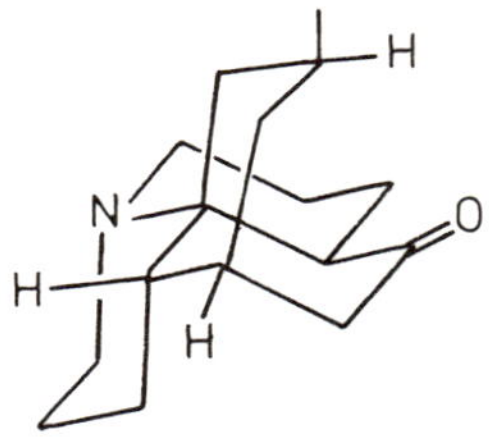

The construction of the intricate Lycopodine framework is achieved by Nature[1] in a simple and fascinating manner from Lysine and Aceto-acetyl-CoA.

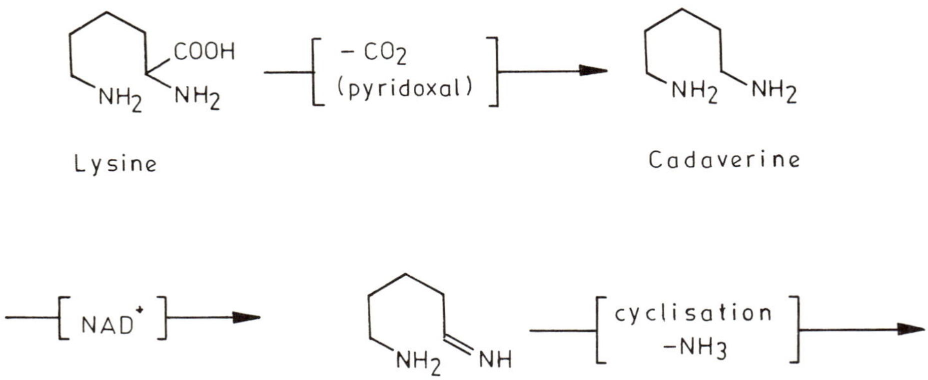

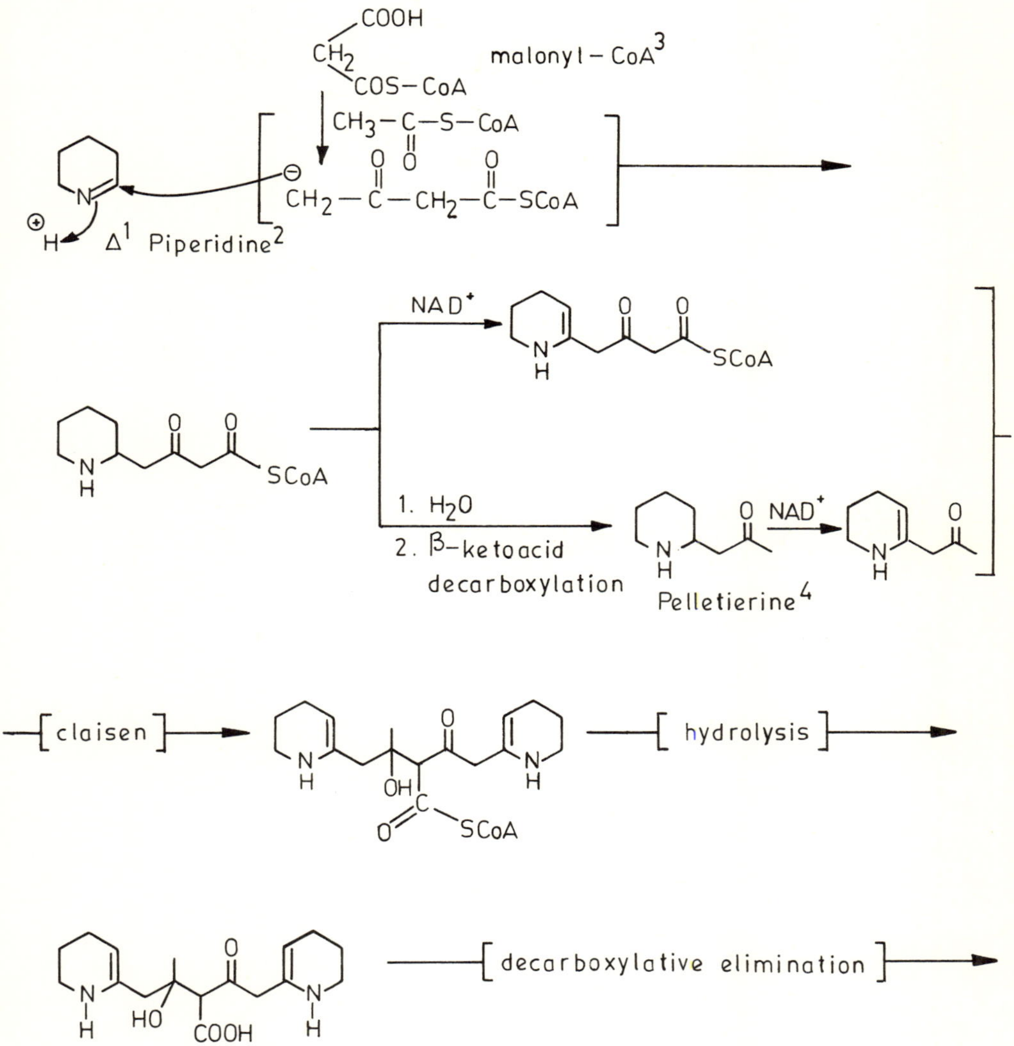
COOH
CH2
COS−CoA
malonyl−CoA3
CH3−C−S−CoA
O
CH2−C−CH2−C−SCoA
Δ1 Piperidine2
NAD+
SCoA
1. H2O
2. β−ketoacid decarboxylation
Pelletierine4
claisen
hydrolysis
OH
COOH
decarboxylative elimination

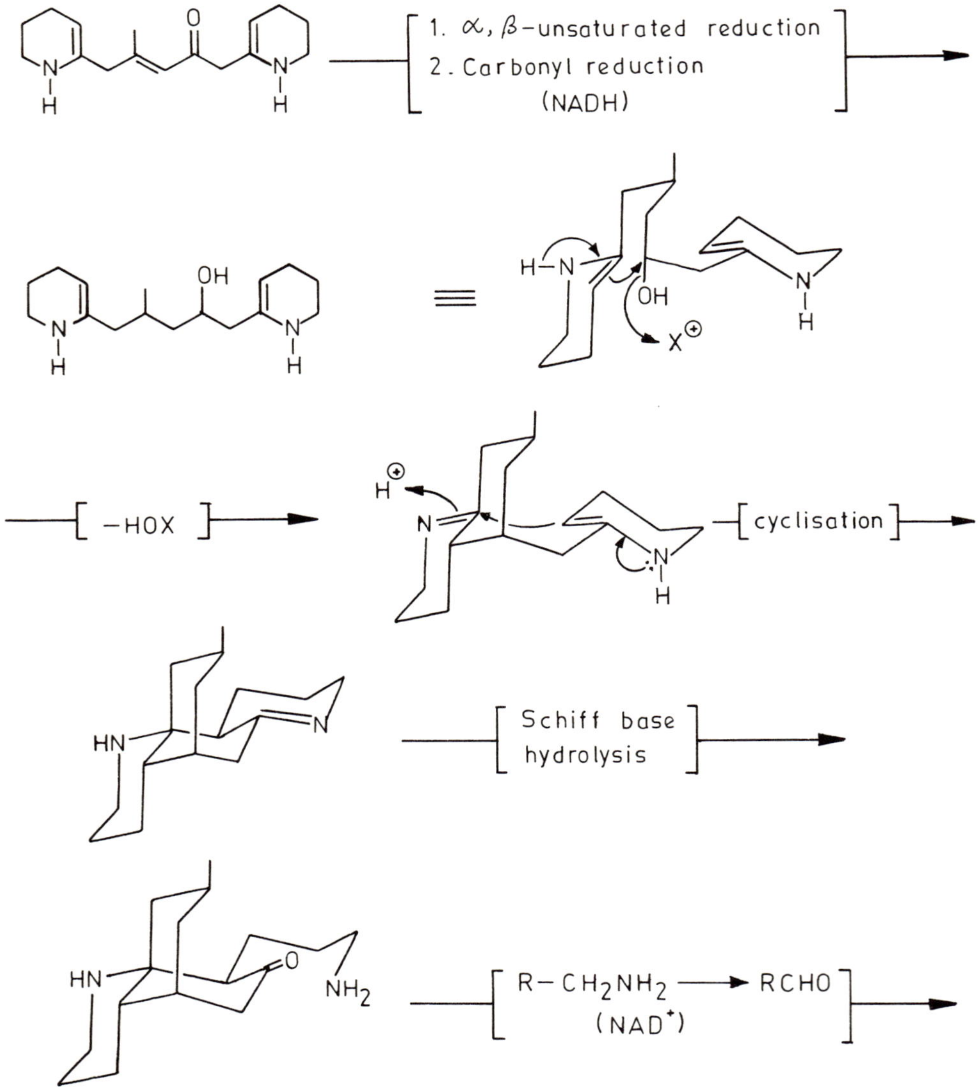

O
N
H
N
H
1. α, β-unsaturated reduction
2. Carbonyl reduction
(NADH)
OH
N
H
N
H
H–N
OH
X⊕
N
H
–HOX
H⊕
N
N
H
cyclisation
HN
N
Schiff base
hydrolysis
HN
O
NH2
R–CH2NH2 → RCHO
(NAD+)

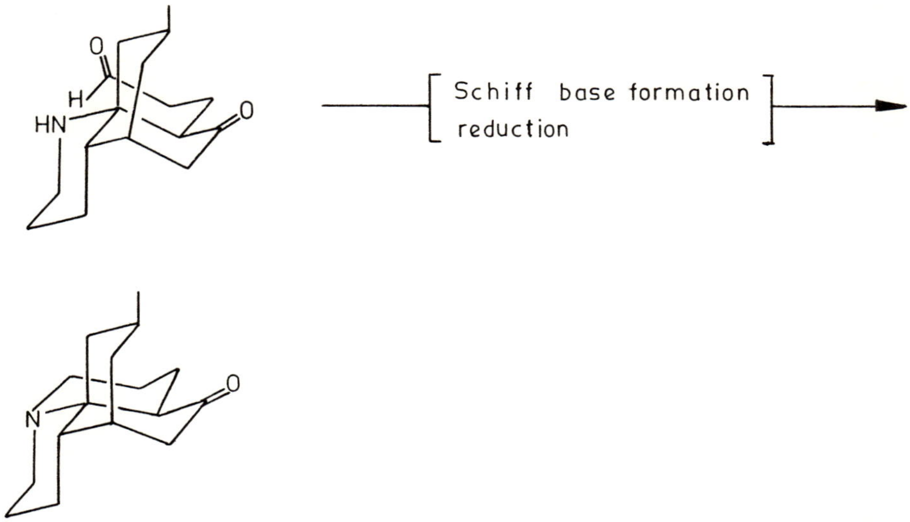

Lycopodine

---

1. R. N. Gupta, M. Castillo, D. B. Maclean, I. D. Spenser and J. T. Wrobel, J. Amer. Chem. Soc., 90, 1360 (1968); M. Castillo, R. N. Gupta, D. B. Maclean and I. D. Spenser, Can. J. Chem., 48, 1893 (1970); M. Castillo, R. N. Gupta, Y. K. Ho, D. B. Maclean and I. D. Spenser, J. Amer. Chem. Soc., 92, 1074 (1970); M. Castillo, R. N. Gupta, Y. K. Ho, D. B. Maclean and I. D. Spenser, Can. J. Chem., 48, 2911 (1970); Y. K. Ho, R. N. Gupta, D. B. Maclean and I. D. Spenser, Can. J. Chem., 49, 3352 (1971); J. C. Braekman, R. N. Gupta, D. B. Maclean and I. D. Spenser, Can. J. Chem., 50, 2591 (1972); W. D. Marshall, T. T. Nguyen, D. B. Maclean and I. D. Spenser, Can. J. Chem., 53, 41 (1975).

2. Experiments with labelled Δ^1 piperidine has shown the following distribution (Ref. 1):

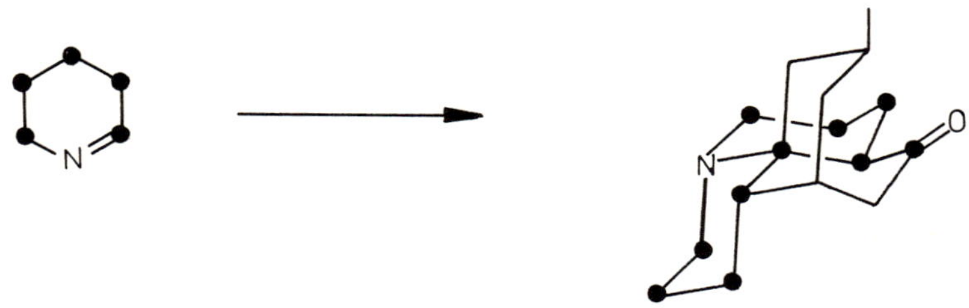

3. Malonic acid serves as a specific precursor of the non-lysine derived segments of lycopodine (Ref. 1).

4. Careful labelled studies have proved that pelletierine itself cannot entirely generate the lycopodine skeleton (Ref. 1):

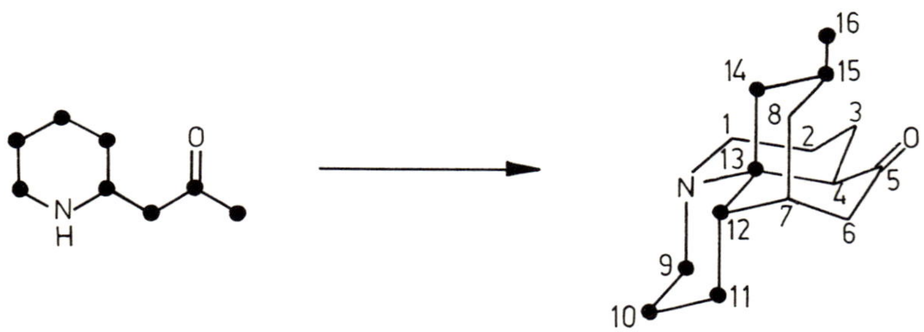

LYSERGIC ACID

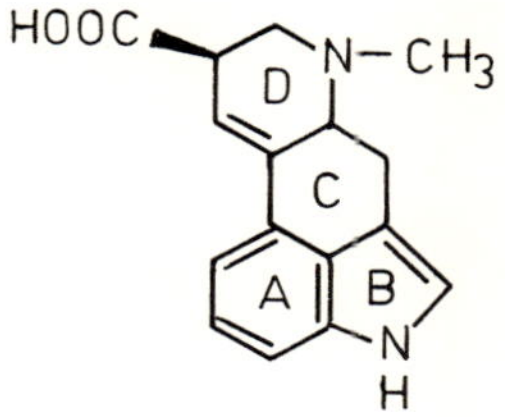

An unusual cyclisation and an unanticipated double isomerisation characterises the biosynthesis[1] of Lysergic and from tryptophan and mevalonic acid[2]!

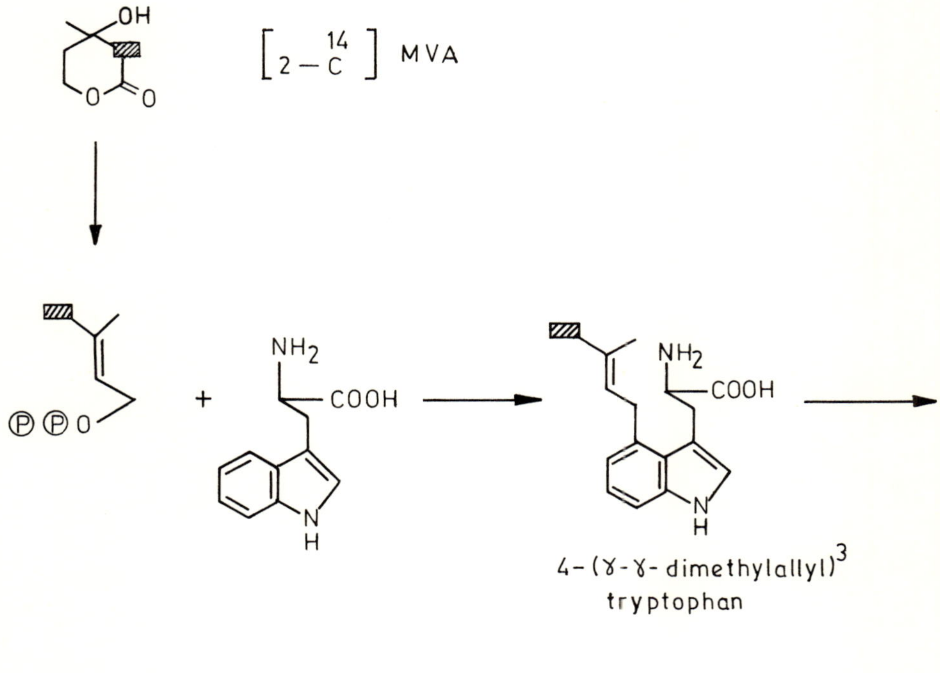

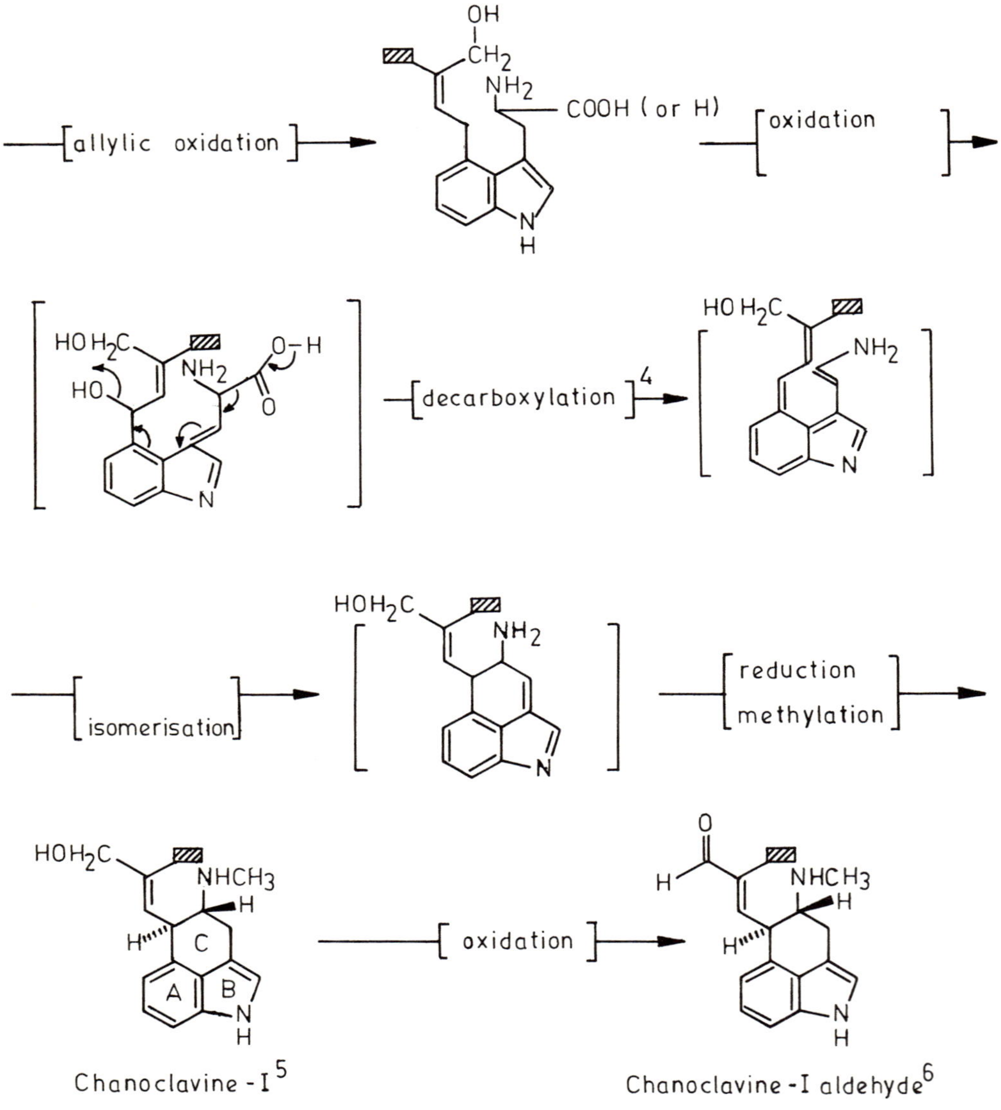
allylic oxidation
OH
CH2
NH2
COOH (or H)
N
H
oxidation
HOH2C
O-H
NH2
HO
O
N
decarboxylation 4
HOH2C
NH2
N
isomerisation
HOH2C
NH2
N
reduction
methylation
HOH2C
NHCH3
H
H
C
A
B
N
H
Chanoclavine - I 5
oxidation
O
H
NHCH3
H
H
N
H
Chanoclavine - I aldehyde 6

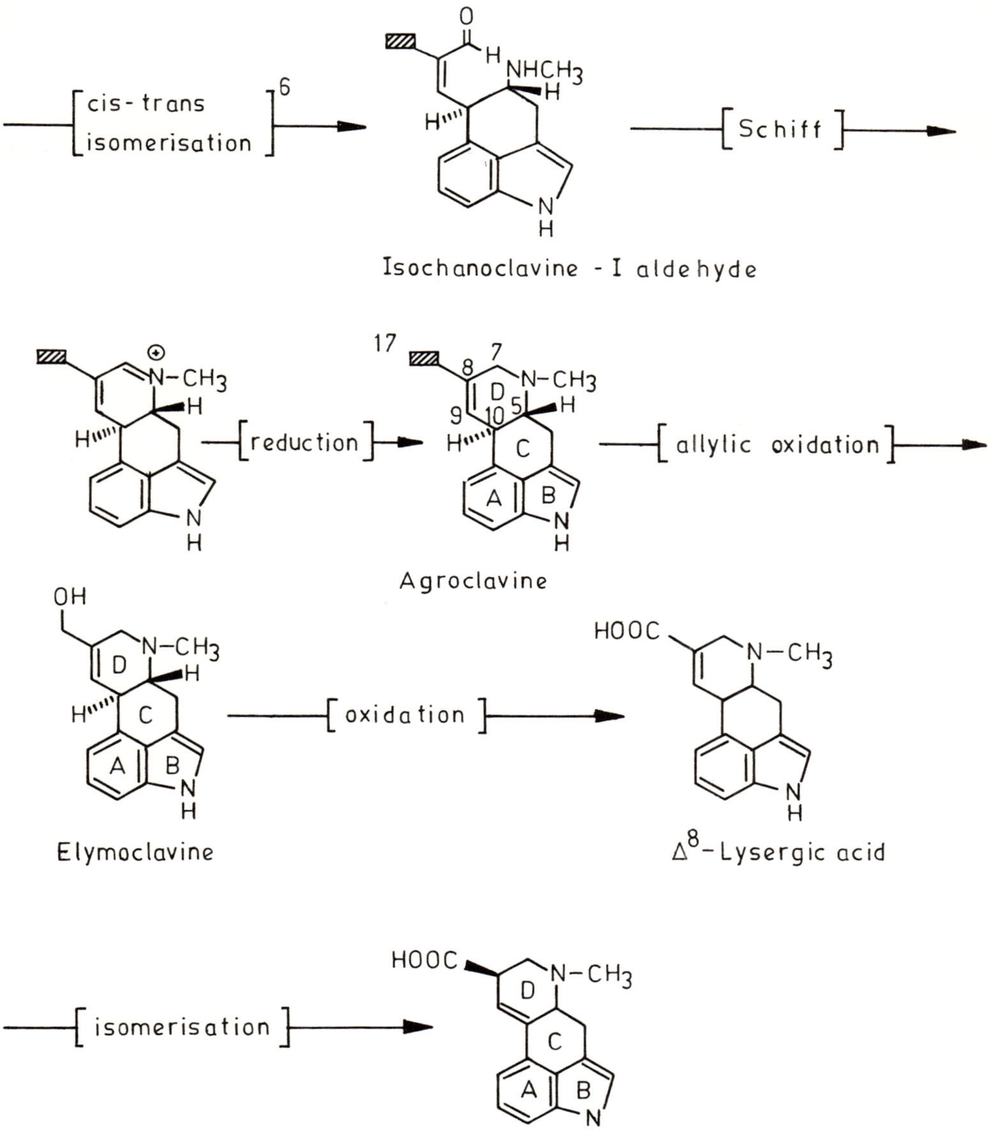
cis-trans isomerisation
6
O
H
NHCH3
H
H
Schiff
Isochanoclavine - I aldehyde
N-CH3
H
H
reduction
17
8
7
N-CH3
D
9
10
5
H
H
C
A
B
N
H
allylic oxidation
Agroclavine
OH
N-CH3
D
H
H
C
A
B
N
H
oxidation
HOOC
N-CH3
N
H
Elymoclavine
Δ8-Lysergic acid
isomerisation
HOOC
N-CH3
D
C
A
B
N
H
Lysergic acid

1. H. G. Floss, Tcheng-Lin, C. J. Chang, B. Naidoo, G. E. Blair, C. I. Abou-Chaar and J. M. Cassady, J. Amer. Chem. Soc., 96, 1898 (1974) and references cited therein; A. I. Scott, in 'Alkaloids', MTP International Review of Science, Organic Chemistry Series one, Vol. 9, (D. H. Hey and K. F. Wiesner, editors), p. 105, Butterworths, University Park Press, 1973.

2. Mevalonic acid is incorporated via γ,γ-dimethylallylpyrophosphate (H. Plieninger, H. Immel and L. Volkl, Ann., 706, 223 (1967)).

3. 4-(γ,γ-dimethylallyl) tryptophan is a known metabolite of ergot fungus under conditions of inhibition of alkaloid synthesis (J. E. Robbers and H. G. Floss, Tetrahedron Letters, 1857 (1969); S. Aguvell and J. E. Lindgreu, Tetrahedron Letters, 5127 (1968); J. E. Robbers and H. G. Floss, Arch. Bioch. Biophys. 126, 967 (1968)). Intact incorporation of this compound into ergot alkaloids has been demonstrated (ref. 2).

4. The decarboxylation precedes cyclisation since with [2-$^{3}H_{1}$] tryptophan, the label is retained in the product.

5. Experiments with [2-^{14}C]MVA showed that the formation of chanoclavine I involves a cis-trans isomerisation at the allylic double bond. The initial hydroxylation takes place at the unlabelled cis (cis to the C-9 substituent) methyl group, which later becomes the trans-hydroxy-methyl group of chanoclavine-I. Chanoclavine I has been successfully incorporated into agroclavine and elymoclavine in ergot fungus (H. G. Floss, V. Hornemann, N. Schilling, K. Kelley, D. Groger, and D. Ergé, J. Amer. Chem. Soc., 90, 6500 (1968).

6. Double labelling experiments with [7-^{14}C-9-^{3}H]-chanoclavine I and 3',4-d_5-MVA indicated that the cyclisation of chanoclavine I to tetracyclic ergolines involves an intermolecular transfer of the C-9 proton and the loss of one methylene hydrogen from the -CH_2OH group of chanoclavine I. Confirmation that chanoclavine I-aldehyde is a precursor was obtained by feeding [7-^{14}C-9-^{3}H] aldehyde and isolating elymoclavine with 40% incorporation (B. Naidoo, J. M. Cassady, G. E. Blair and A. G. Floss, Chem. Comm., 671 (1970)). The mechanism of inversion of chanoclavine-I aldehyde is as follows

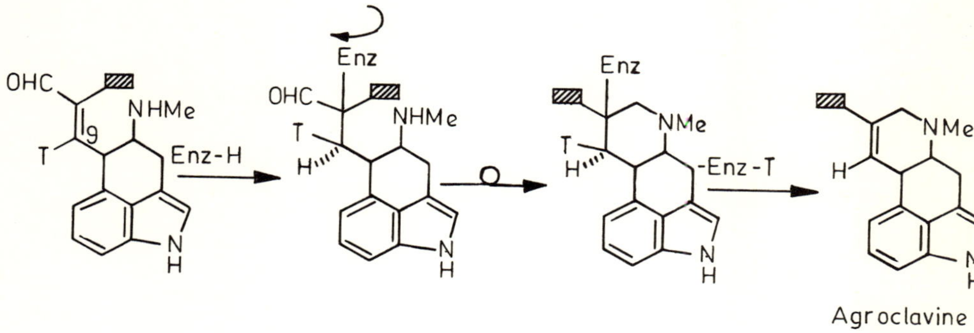
OHC
NHMe
9
T
Enz-H
Enz
OHC
T
H
NHMe
O
Enz
NMe
T
H
-Enz-T
NMe
H
N
H
Agroclavine

S-ADENOSYL METHIONINE

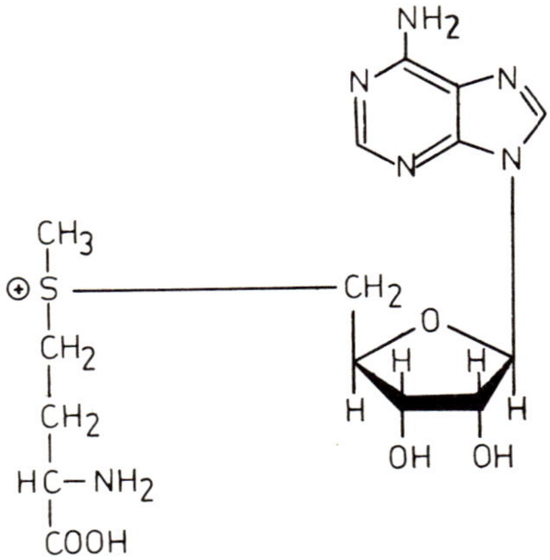

The biosynthesis of S-adenosylmethionine - the ubiquitous methylating agent - illustrates the manoeuvres that the living systems have evolved from profound experience, highlighting the fact that in assemblies that make various types of molecules, the longer route can be not only prettier but also more easy![1]

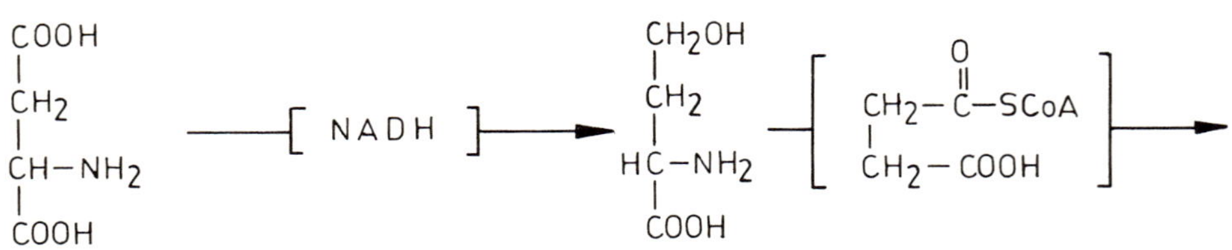

Aspartic acid

Homoserine

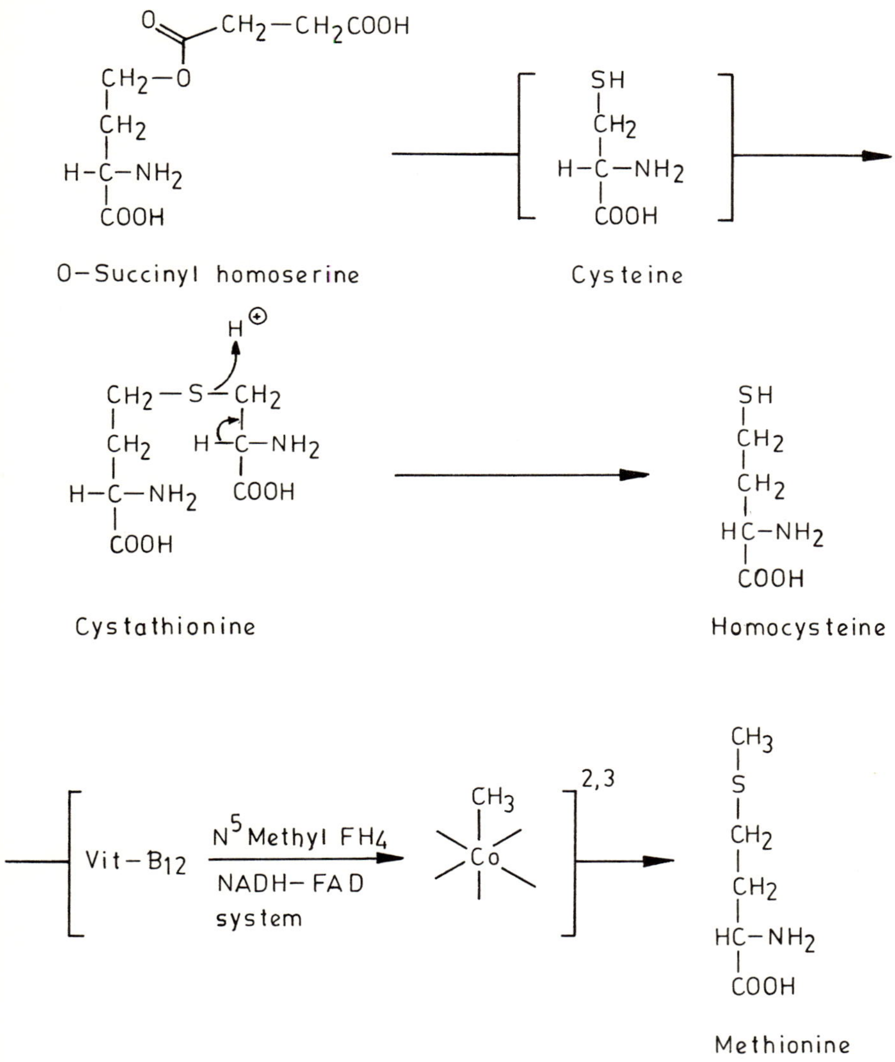
O
CH2—CH2COOH
CH2—O
CH2
H—C—NH2
COOH
O—Succinyl homoserine
SH
CH2
H—C—NH2
COOH
Cysteine
H⊕
CH2—S—CH2
CH2
H—C—NH2
COOH
H—C—NH2
COOH
Cystathionine
SH
CH2
CH2
HC—NH2
COOH
Homocysteine
Vit—B12
N5 Methyl FH4
NADH—FAD
system
CH3
Co
2,3
CH3
S
CH2
CH2
HC—NH2
COOH
Methionine

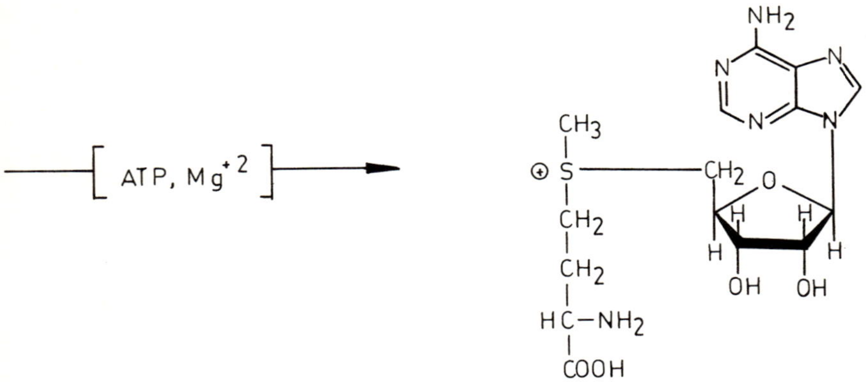

S– adenosyl methionine

---

1. See 'Secondary Metabolism in Plants and Animals', Ed. M. Luckner, Chapman and Hall Ltd., London, 1972, p. 233.

2. The $Co^{III} \rightarrow Co^{I}$ system that is possible with Vitamin-B_{12} is most likely involved in this change:

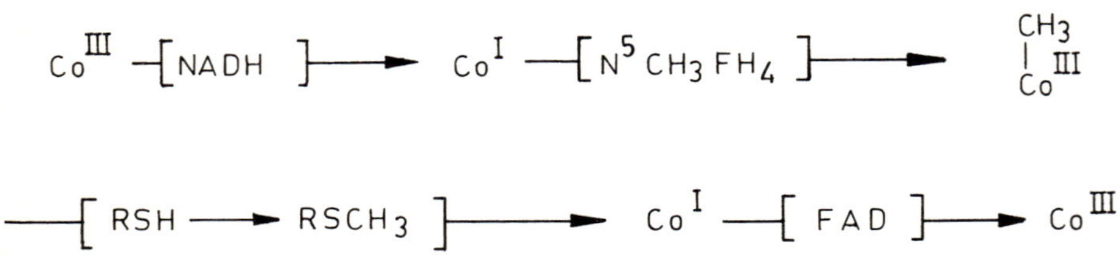

3. Interestingly, this methyl transfer is catalysed by s-adenosyl methionine!

MEVALONIC ACID

⇅

MEVALONIC LACTONE

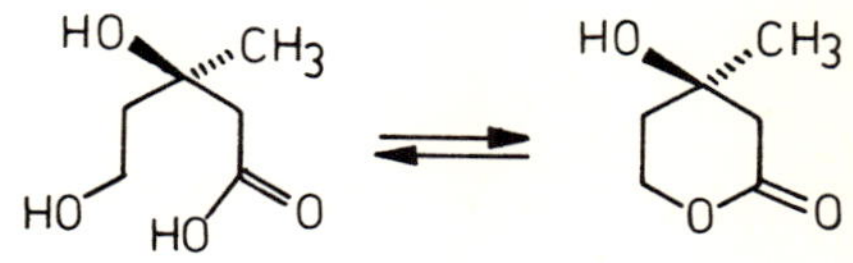

Mevalonic acid is one of the more favoured synthons in the Nature's art in biosynthesis. It is a precursor to an innumerable array of diverse types of bio-molecules that are intimately associated with life processes. The biosynthesis of Mevalonic acid is straightforward and involves the preferential reduction of one of the carboxyl groups of a glutaric acid unit![1,2,3]

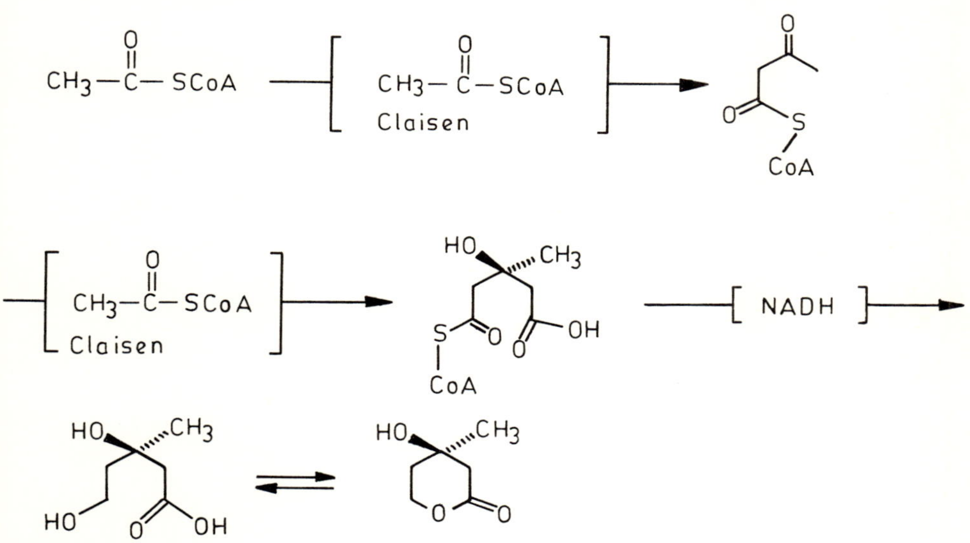

Mevalonic acid   Mevalonic lactone

1. For recent synthesis of (R)- and (S)- Mevalonic acids, see, F. C. Huang, L. F. Hsu Lee, R. S. D. Mittal, P. R. Ravi Kumar, J. A. Chan, C. J. Sih, E. Caspi and C. R. Eck, J. Amer. Chem. Soc., 97, 4144 (1975).

2. For a review of the topic, see, J. W. Cornforth and R. H. Cornforth in 'Natural Substances Formed Biologically from Mevalonic Acid', Ed., T. W. Goodwin, Academic Press, New York, N.Y. 1970, p. 5.

3. For Biosynthesis see, R. B. Clayton, Quart. Rev., 19, 168 (1965); J. W. Cornforth, Quart. Rev., 23, 125 (1969) and references cited therein.

MORPHINE

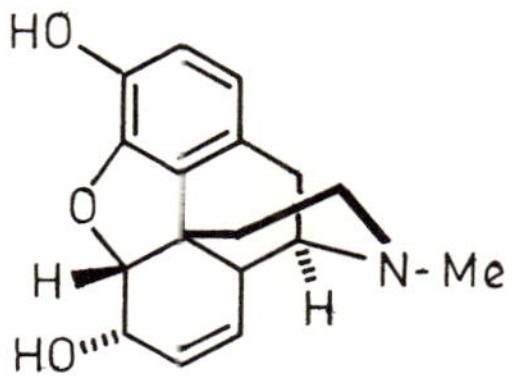

The biogenetic relationship between 1-benzylisoquinoline system and morphine recognized by Robinson - as early as 1925![1] - has now been established.[2] The chemical simulation of the morphine bio-synthesis has also been achieved.[3]

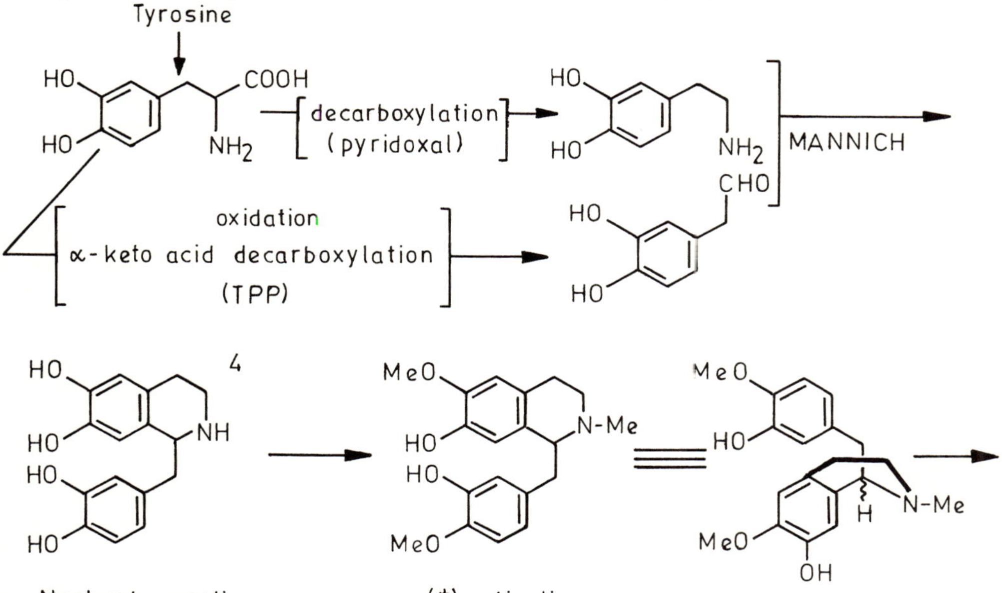

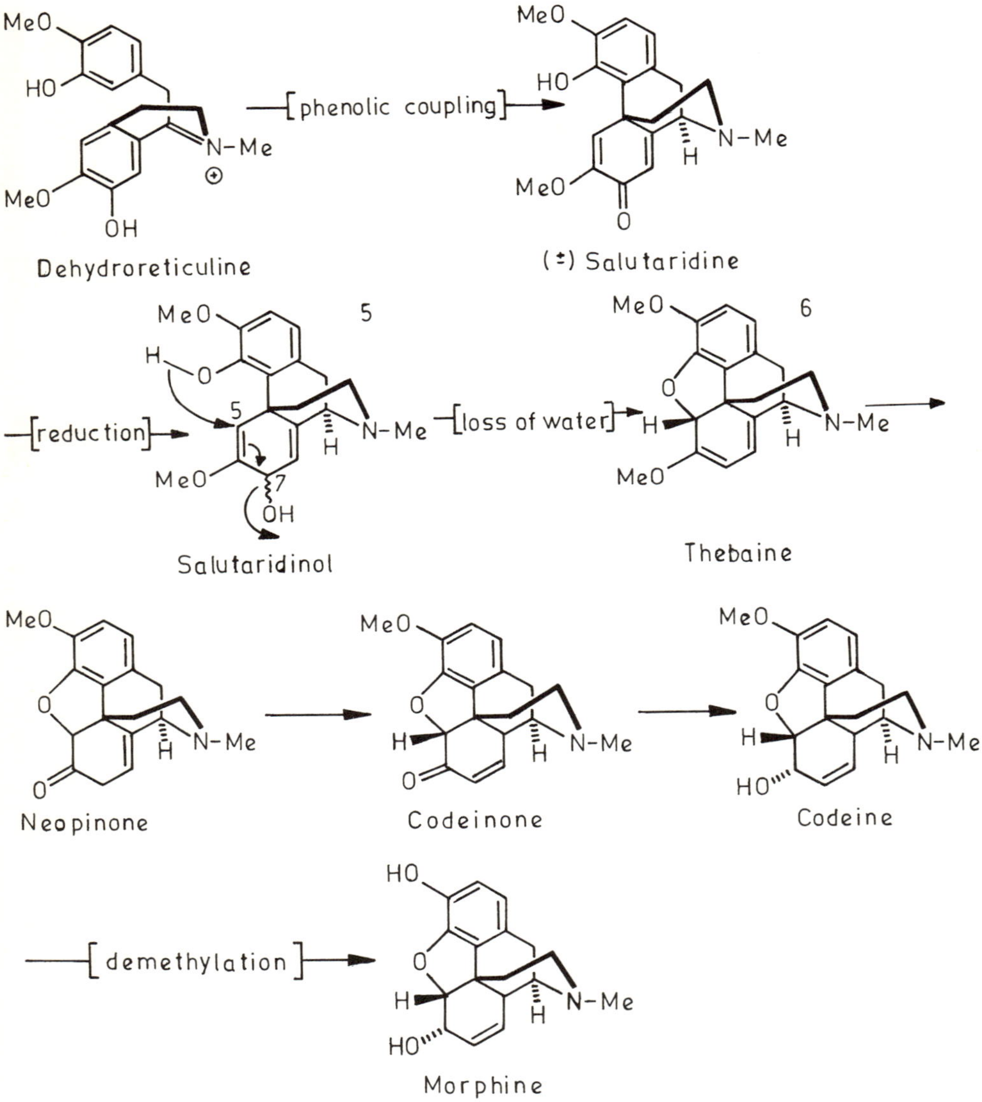
Dehydroreticuline
phenolic coupling
(±) Salutaridine
reduction
5
Salutaridinol
loss of water
6
Thebaine
Neopinone
Codeinone
Codeine
demethylation
Morphine

1. J. M. Gull and R. Robinson, Mem. Proc. Manchester, Lit. Phil. Soc., 69, 79 (1925).

2. A. R. Battersby and R. Binks, Proc. Chem. Soc., 360 (1960); A. R. Battersby, R. Binks, R. J. Francis, D. J. McCaldin and H. Ramuz, J. Chem. Soc., 3600 (1964).

3. D. H. R. Barton, G. W. Kirby, W. Steglich and G. M. Thomas, Proc. Chem. Soc., 203 (1963).

4. Suitable biological methylation of norlaudanosoline generates reticuline which by way of the dehydroderivative and then by the ortho-para diradical coupling mode (D. H. R. Barton and T. Cohen, Festschr. Arthur. Stoll, 117 (1957) is converted into salutaridine. Labelled reticuline and 1,2-dehydroreticuline gave a high and specific incorporation into morphine in P. somniferum plants (A. R. Battersby, D. M. Foulkes and R. Binks, J. Chem. Soc , 3323 (1965)).

5. The 7β-hydroxyl configuration in salutaridinol is unfavourable for an intramolecular SN_2' displacement. Consequently, in the event the process is concerted, the 7-centre most probably undergoes an enzyme mediated inversion which would make the SN_2' leading to thebaine possible (D. H. R. Barton, D. S. Bhakuni, R. James and G. W. Kirby, J. Chem. Soc., C, 128 (1967)).

6. The thebaine → morphine transformation via codeinone and codeine has been established (A. R. Battersby, J. A. Martin and E. Brockmann-Hanssen, J. Chem. Soc., C, 1785 (1967); G. Blaschke, H. I. Parker and H. Rapoport, J. Amer. Chem. Soc., 89, 1540 (1967)). In a more detailed study of the biosynthetic conversion of thebaine to codeine by P. somniferum, it has been established (H. I. Parker, G. Blaschke and H. Rapoport, J. Amer. Chem. Soc., 94, 1276 (1972)) that thebaine is converted to codeine by initial demethylation to neopinone followed by rearrangement to codeinone which is then reduced to codeine. The observed (D. H. R. Barton, G. W. Kirby, W. Steglich, G. M. Thomas, A. R. Battersby, T. A. Dobson and H. Ramuz, J. Chem. Soc., 2423 (1965)) loss of tritium at C-7 during the conversion of dual-labeled thebaine to codeine also may be explained by this route if a nonspecific addition or removal of hydrogen at C-7 occurs during conversion to neopinone and codeinone.

PATULIN

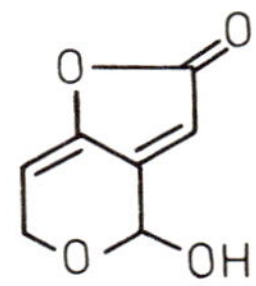

In Patulin, an over abundance of functional groups adorn a slender carbon frame. The creation of this mold metabolite by Nature[1] is a typical illustration of the methodology involved in the metabolism of electron rich aromatic systems. These are readily cleaved and remoulded to provide a variety of structures.

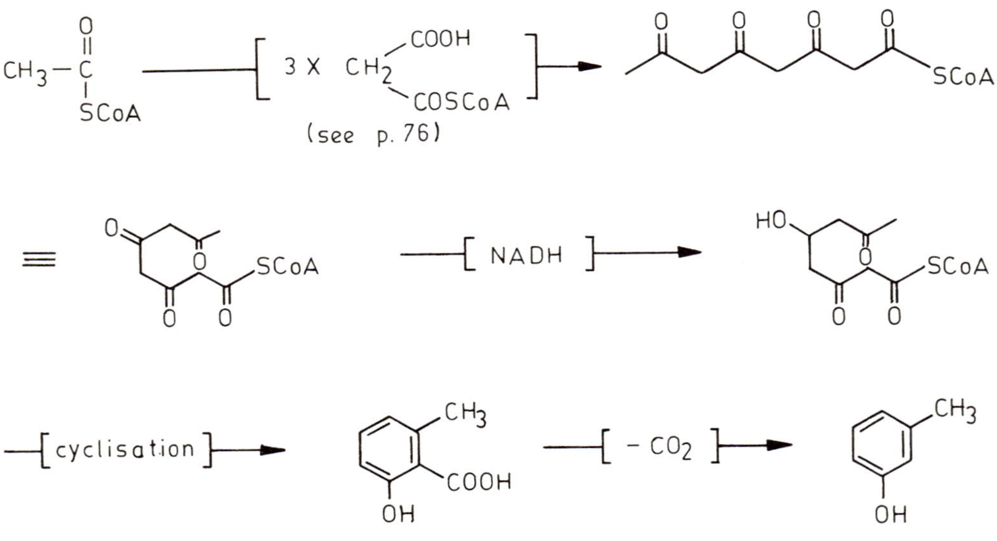

6-methyl salicylic acid

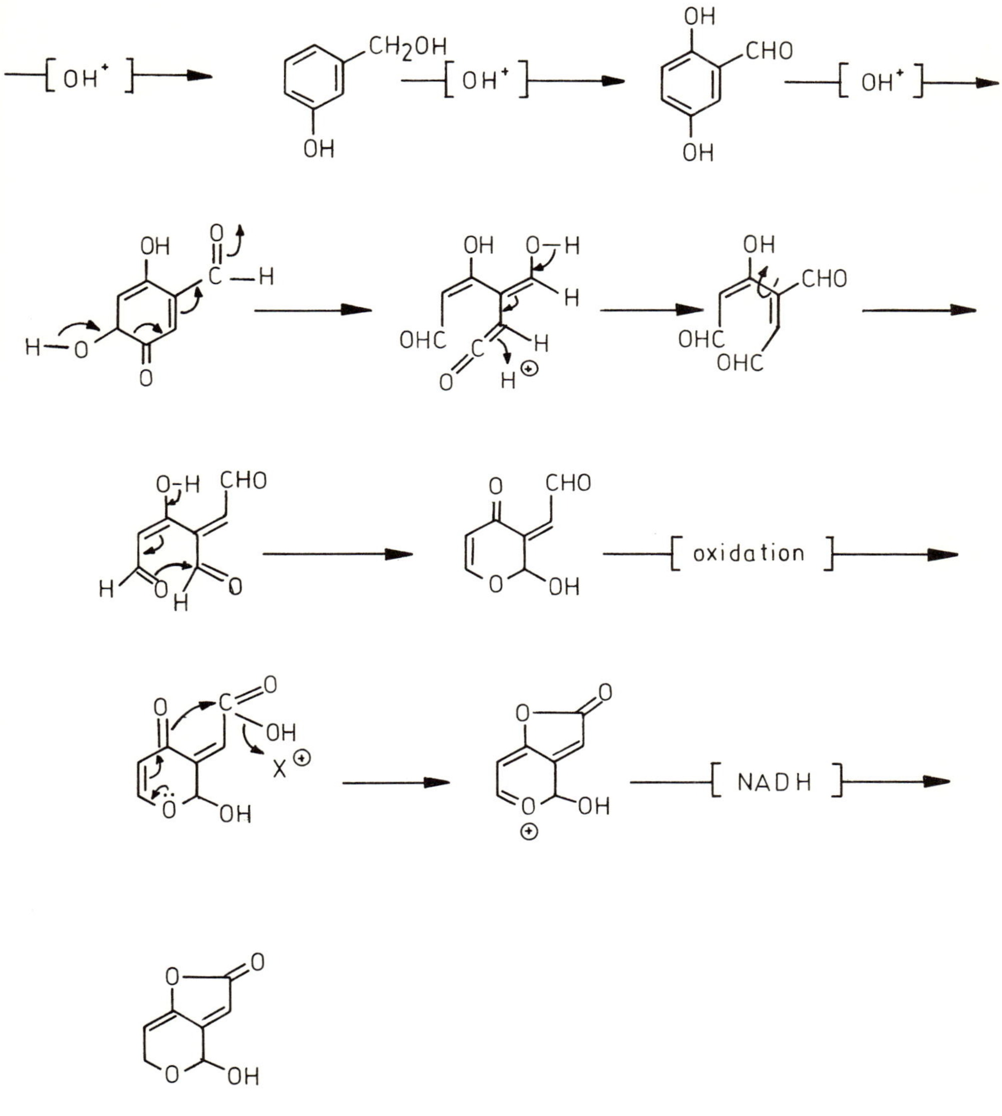
[OH+]
CH2OH
OH
[OH+]
OH
CHO
OH
[OH+]
OH
O
C —H
H—O
O
OH
O—H
H
OHC
H
O
H⊕
OH
CHO
OHC
OHC
O—H
CHO
H
O
H
O
O
CHO
O
OH
[oxidation]
O
C
O
OH
X⊕
O
OH
O
O
O
OH
⊕
[NADH]
O
O
O
OH

Patulin

1. A. I. Scott, L. Zamir, G. T. Phillips and M. Yalpani, Bioorg. Chem., 2, 124 (1973); A. I. Scott and M. Yalpani, Chem. Comm., 945 (1967); P. L. Forrester and G. M. Gaucher, Biochemistry, 11, 1102 (1972); A. I. Scott and L. Beadling, Bioorg. Chem., 3, 281 (1974).

PHENYL ALANINE, TYROSINE

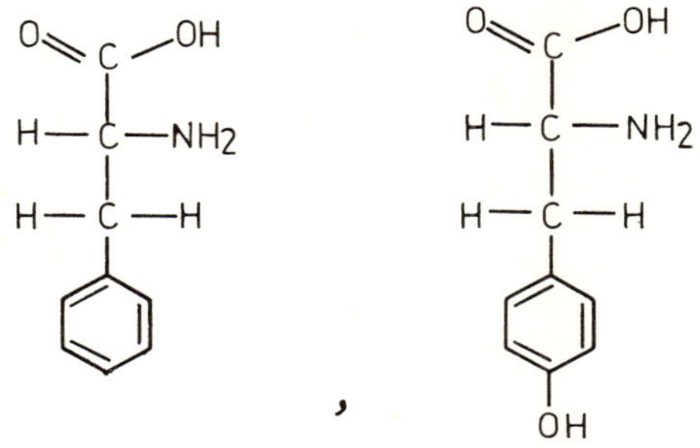

Phenylalanine, Tyrosine

The creation of Phenylalanine and Tyrosine illustrates a favourite method for the generation of phenyl group in living system![1] This is achieved in a most deceptively simple manner from recognised intermediates involved in carbohydrate metabolism. The path to Phenylalanine and Tyrosine involves the intermediacy of Shikimic acid that is related to a host of natural products containing a six membered ring unit[2] and involves a fascinating [3,3] shift!

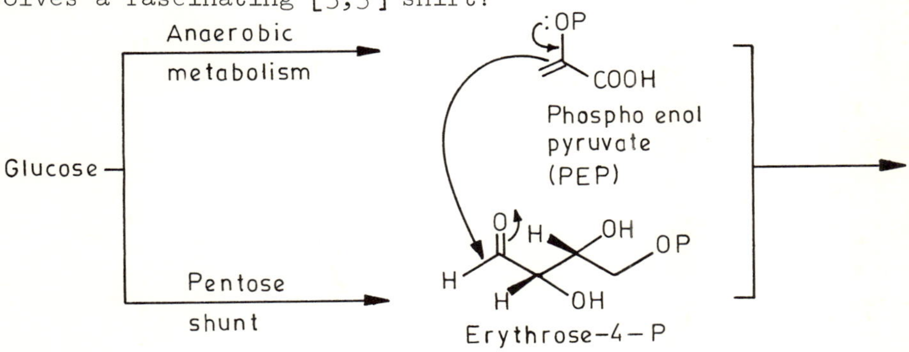

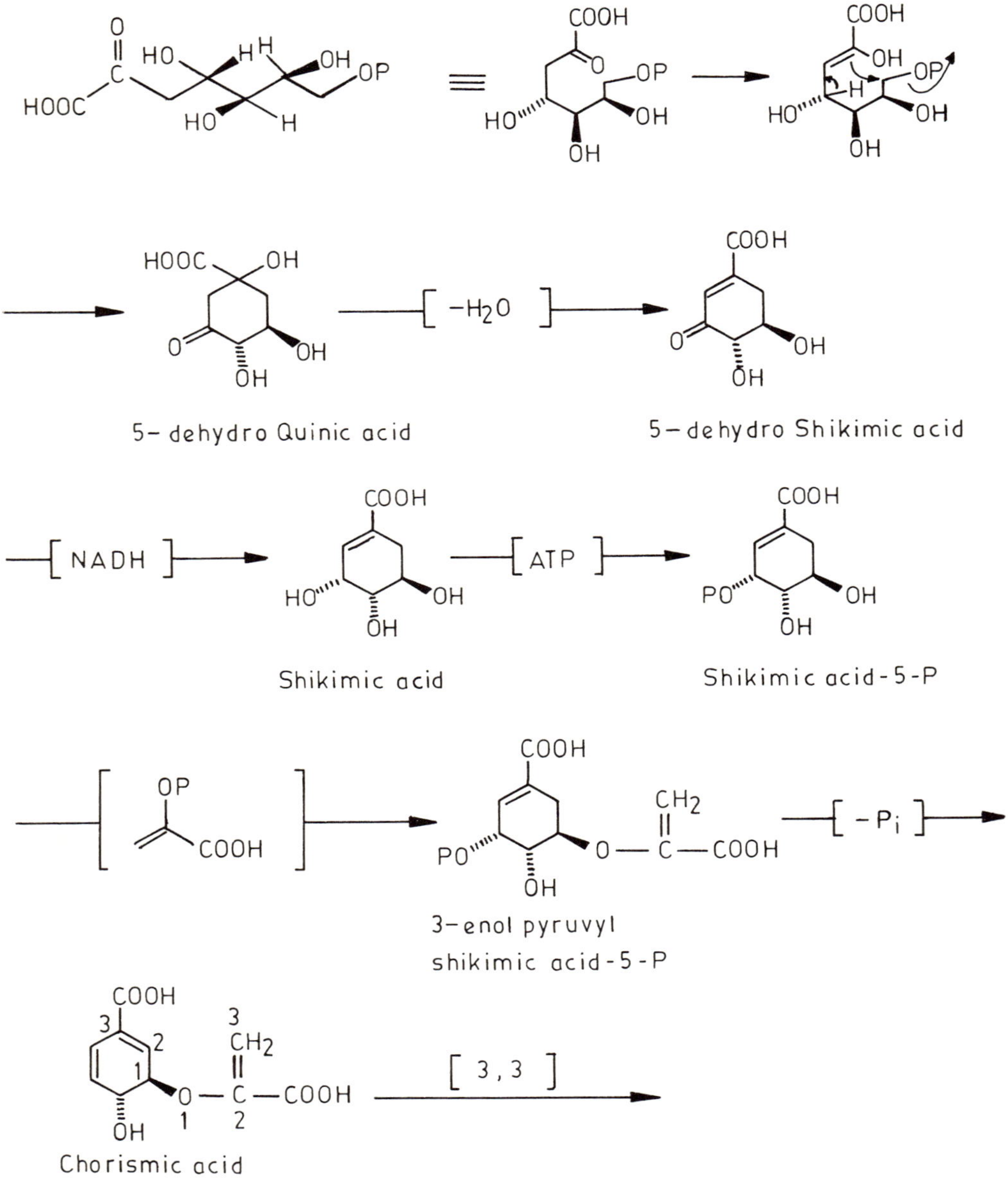

5- dehydro Quinic acid
−H2O
5– dehydro Shikimic acid
NADH
Shikimic acid
ATP
Shikimic acid-5-P
−Pi
3–enol pyruvyl
shikimic acid-5-P
Chorismic acid
3,3

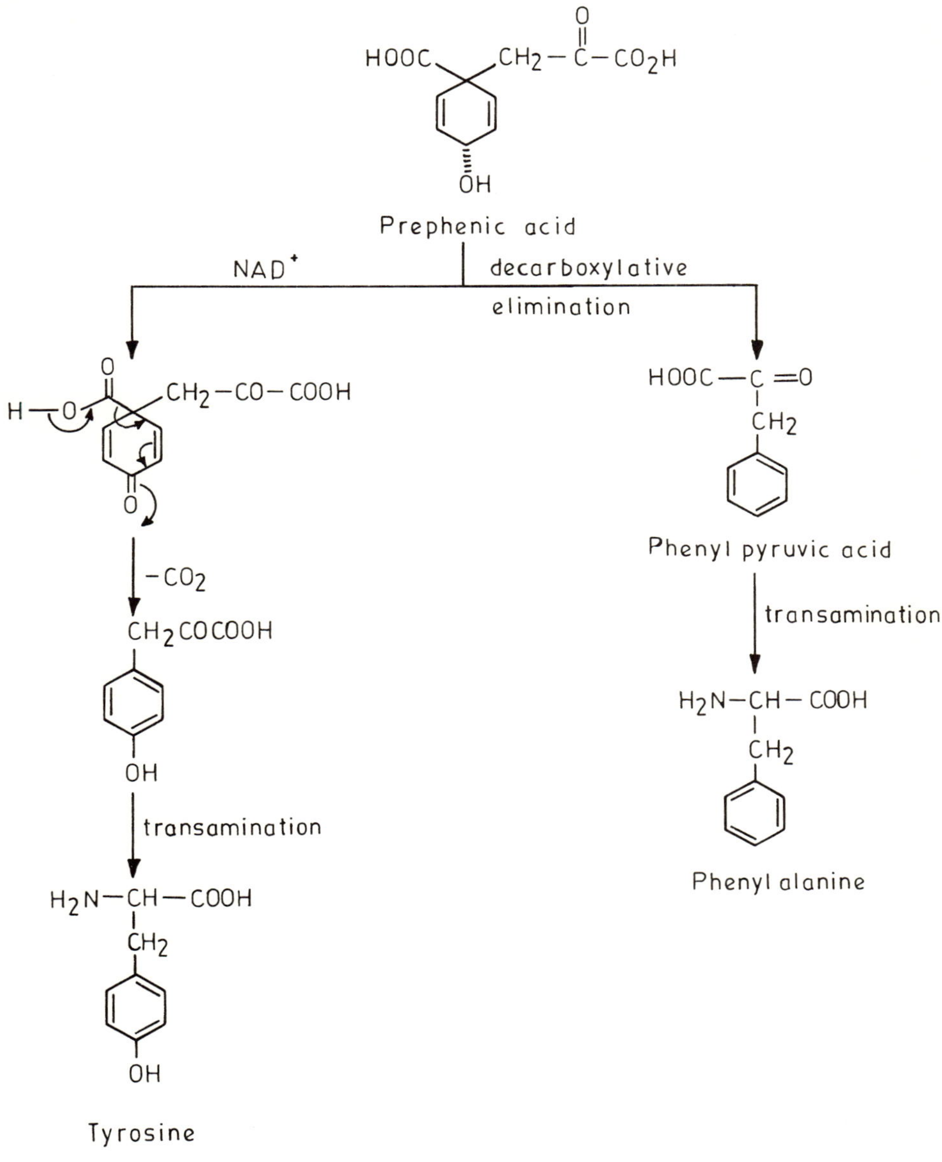
O
HOOC
CH2 – C – CO2H
OH
Prephenic acid
NAD+
decarboxylative
elimination
O
H – O
CH2 – CO – COOH
O
–CO2
CH2COCOOH
OH
transamination
H2N – CH – COOH
CH2
OH
Tyrosine
HOOC – C =O
CH2
Phenyl pyruvic acid
transamination
H2N – CH – COOH
CH2
Phenyl alanine

1. See 'Biological Chemistry', Second Edition, Eds. H. R. Mahler and E. H. Cordes, Harper and Row, New York, 1971, p. 791.

2. B. D. Sprinson, Advances in Carbohydrate Chemistry, 15, 235 (1960).

PROSTAGLANDINS

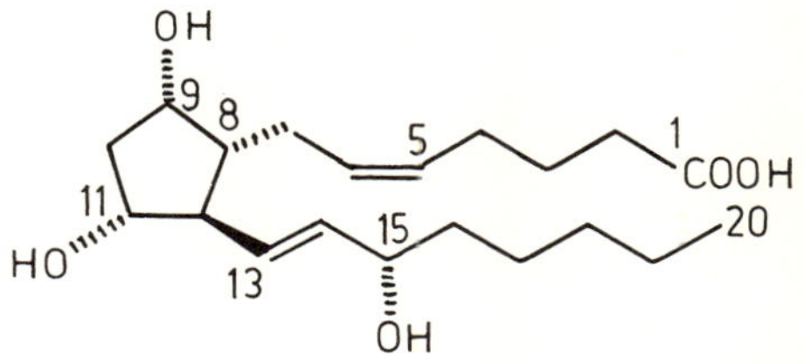

$PGF_{2}\alpha$

Prostaglandins[1]! Rarely, if ever, has so much attention been focussed on any molecular array as Prostaglandins. Even when compared with the two fascinating aspects of Prostaglandins, namely, their synthesis and their profound and diverse biological action, the creation of these molecules by Nature - from appropriate fatty acids[2] - stands out brilliantly in that the smooth formation of the five asymmetric centres from an achiral precursor takes place with the introduction of three oxygen functions in a manner for which there is no parallel either in organic chemistry or in biology[3]!

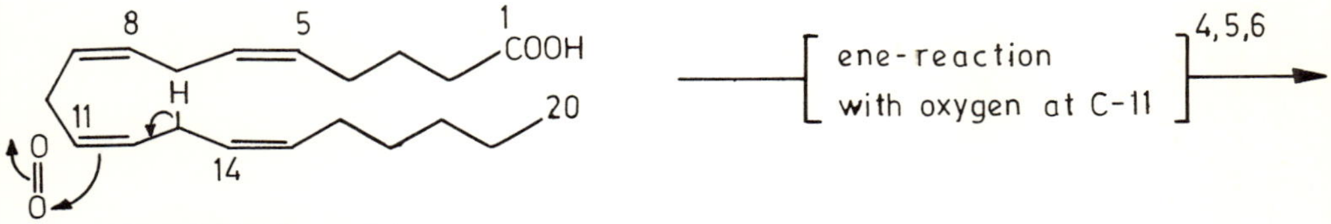

Arachidonic acid

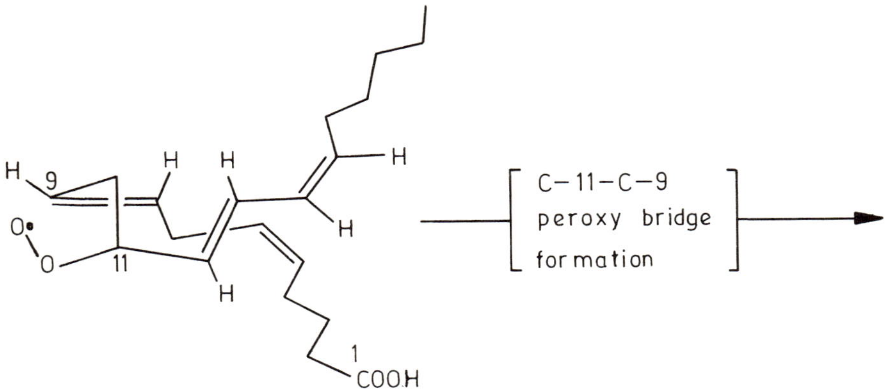
H
9
O•
O
11
H
H
H
H
H
H
1
COOH
C-11-C-9
peroxy bridge
formation

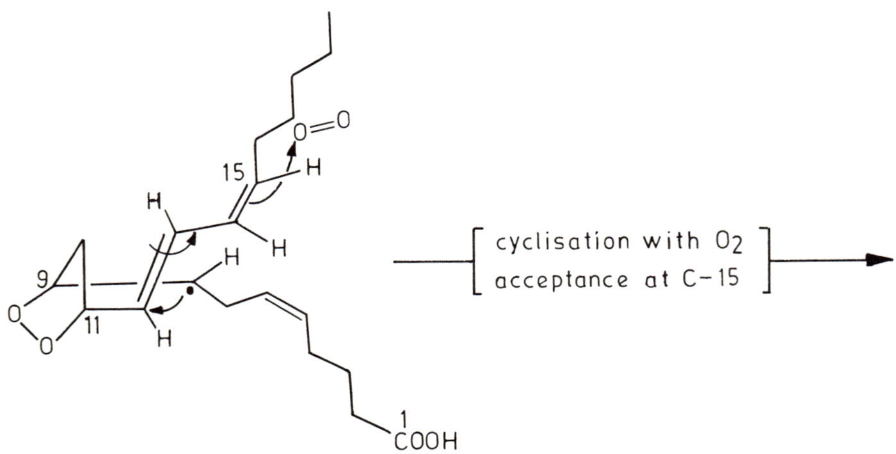
O=O
15
H
H
H
H
9
O
O
11
H
1
COOH
cyclisation with O_2
acceptance at C-15

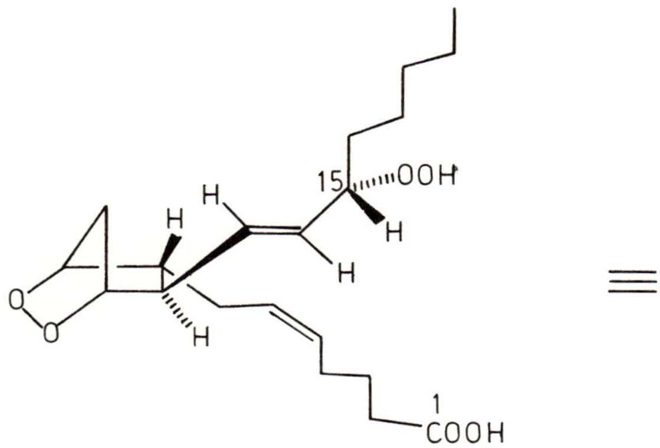

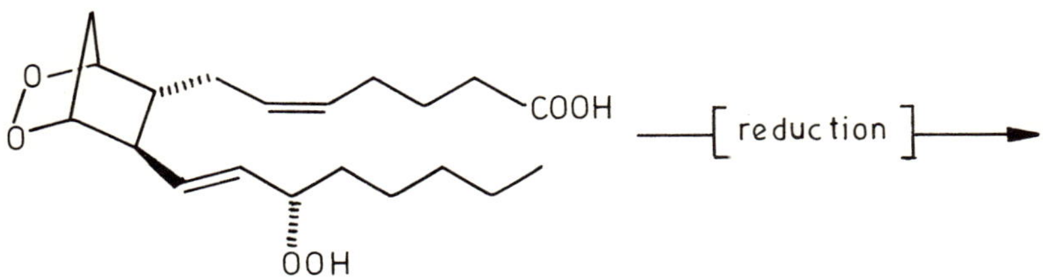

PG-endoperoxide
(PGG_2)[7]

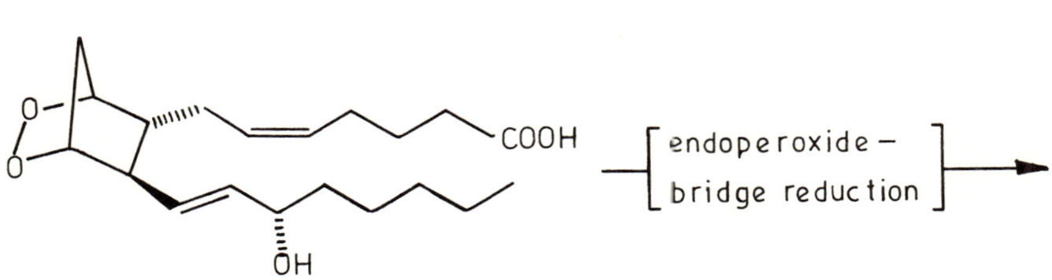

PG-endoperoxide
(PGH_2)[8,9]

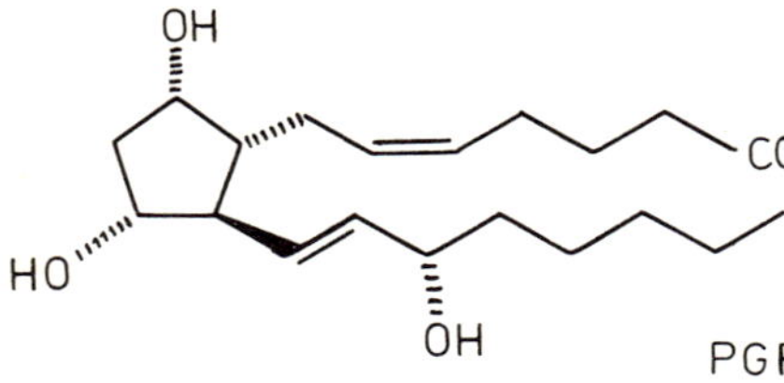

$PGF_2\alpha$

1. All prostaglandins are, in principle, derived from the parent Prostanoic acid.

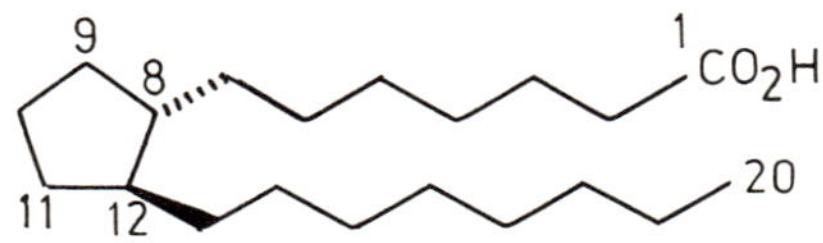

Prostanoic acid

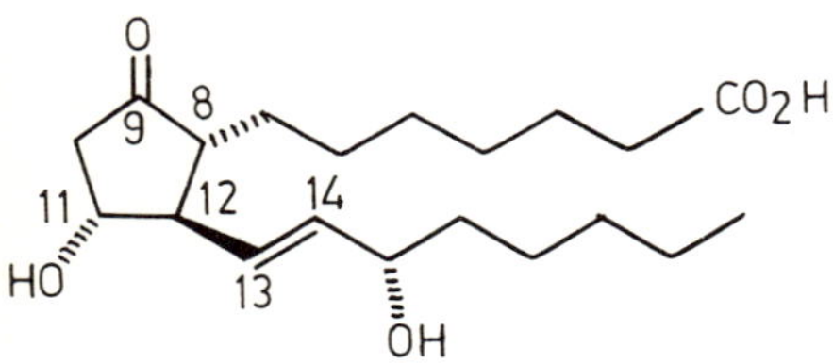

PGE_1

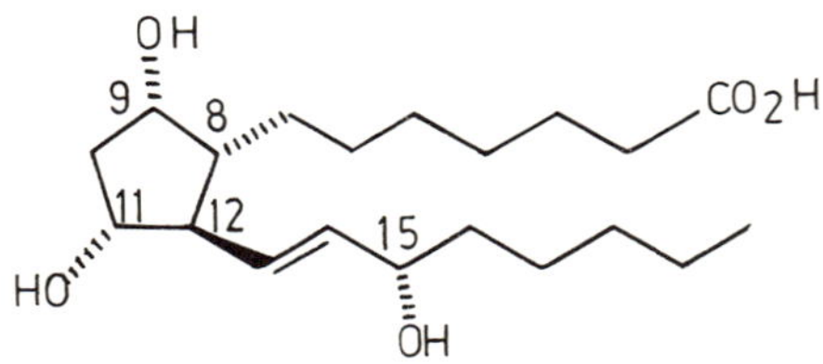

$PGF_1\alpha$

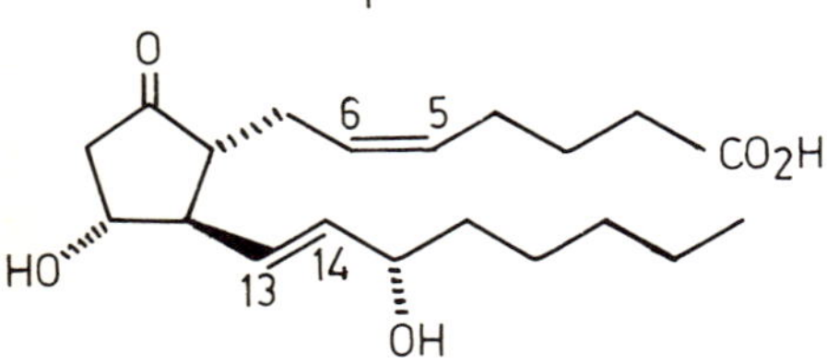

PGE_2

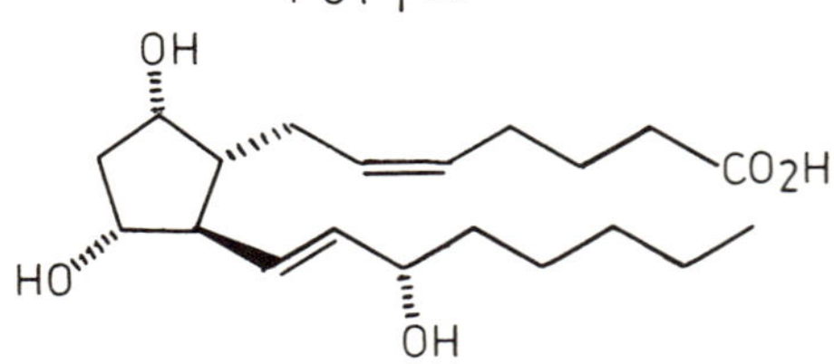

$PGF_2\alpha$

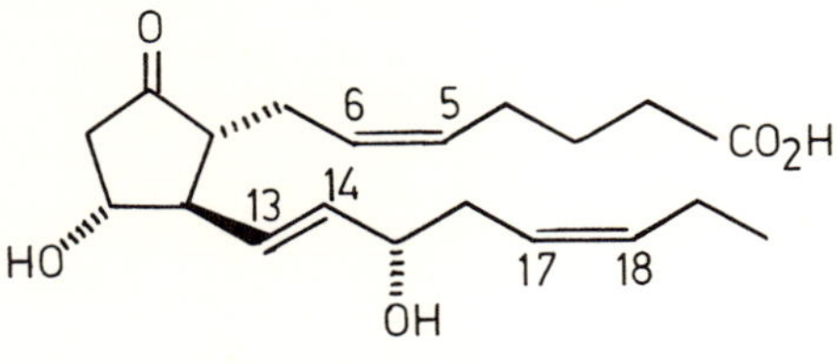

PGE_3

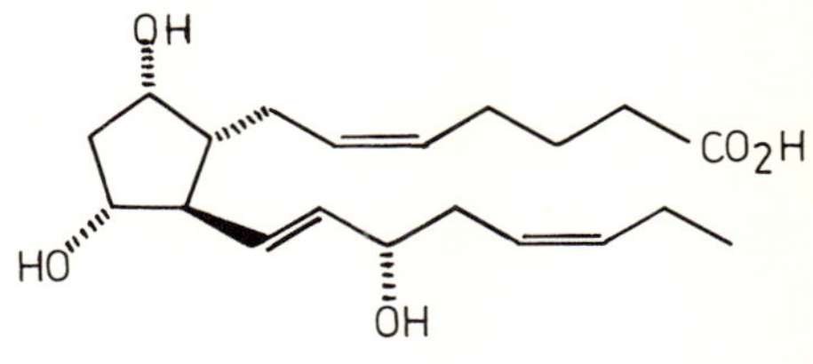

$PGF_{3}\alpha$

2. A minimum requirement for obtaining a prostaglandin from a fatty acid is the presence of at least three cis double bonds in a methylene interrupted chain:

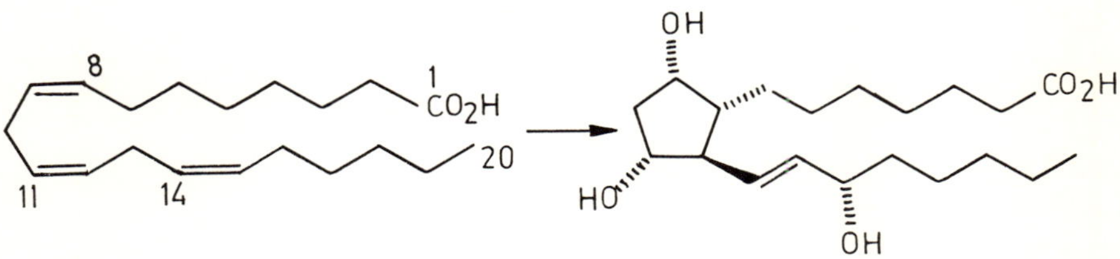

8,11,14-eicosatrienoic acid

$PGF_{1}\alpha$

C_{19} and C_{21} fatty acids fulfilling this minimum requirement also undergo similar cyclisation.

3. D. A. van Dorp, R. K. Beerthuis, D. H. Nugteren and H. Vonkeman, Biochim. Biophys. Acta, 90, 204 (1964); S. Bergstrom, H. Danielsson and B. Samuelsson, Biochim. Biophys. Acta, 90, 207 (1964); E. Anggard and B. Samuelsson, J. Biol. Chem., 240, 3518 (1965).

4. Molecular oxygen is essential for the biosynthesis since the incubations carried out anaerobically do not produce any prostaglandins.

5. The enzyme responsible for the conversion of essential fatty acids to prostaglandins is called PG synthetase. Its presence has been shown in various organs of rat, guinea pig, sheep, bull and humans. The synthetase can be used either as the homogenate or as acetone powder.

6. The isolation of 11-hydroxy-12-trans-14-cis-eicosadienoic acid Ⓐ, from 11,14-eicosadienoic acid by vesicular-glands is in good agreement with the view that the prostaglandin synthesis is initiated by the ene-reaction.

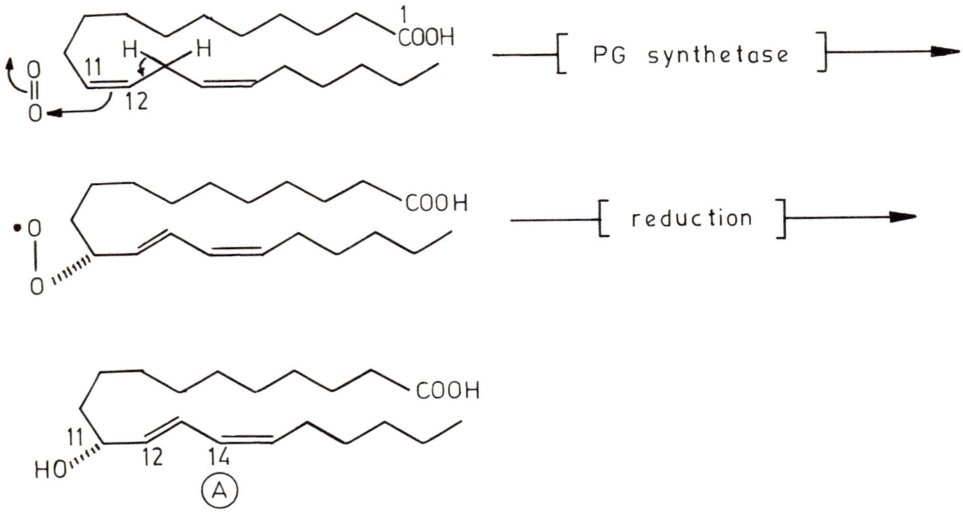

(B. Samuelsson, Fed. Proc., 31, 1442 (1972)).

Additionally, plant lipoxidases, lacking qualities to bring about the intricate transformations associated with the PG-biosynthesis, mainly affect hydroperoxidation at C_{15}, again involving the abstraction of C_{13} hydrogen.

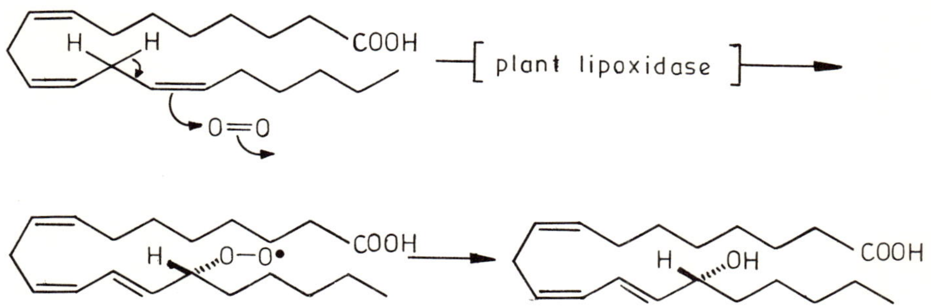

(B. Samuelsson, Prog. Biochem. Pharmac. 3, 59 (1967); D. H. Nugteren, R. K. Beerthuis and D. A. van Dorp, Nobel Symposium 2. Prostaglandins, Eds. S. Bergstrom and B. Samuelsson, Almquist and Wiksell, Stockholm, 1967, p. 45.)

7. The simultaneous formation of $PGF_2\alpha$, 11-dehydro-$PGF_2\alpha$ and PGE_2 is in excellent agreement with the endoperoxide intermediate (PGG_2):

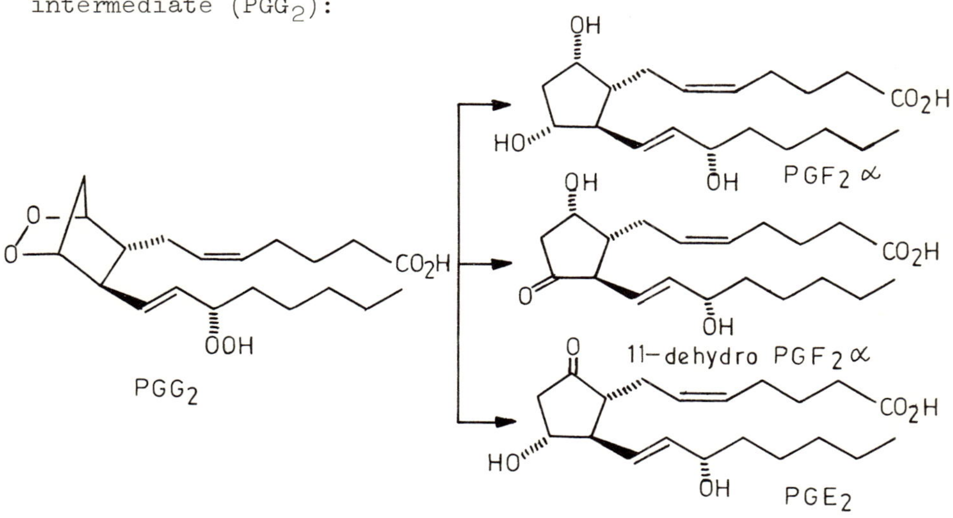

(M. Hamberg and B. Samuelsson, Proc. Nat. Acad. Sci. U.S.A., 70, 899 (1973).)

The intermediacy of PGG_2, first proposed in 1965 (B. Samuelsson, J. Amer. Chem. Soc., 87, 3011 (1965)), has since been confirmed by its isolation and its in situ reactions (M. Hamberg, J. Svensson, T. Wakabayashi and B. Samuelsson, Proc. Nat. Acad. Sci., 71, 345 (1974)).

When PGE_1 was biosynthesised from 8,11,14-eicosatrienoic acid in an atmosphere of $^{18}O_2/^{16}O_2$, the two oxygen functions of the five membered ring contained either two atoms of ^{18}O or two atoms of ^{16}O, but not one atom of ^{16}O and one of ^{18}O. This result showed that the oxygens of the keto and hydroxyl groups at C-9 and C-11 respectively, originated from the same oxygen molecule thus providing good evidence for the endoperoxide intermediate.

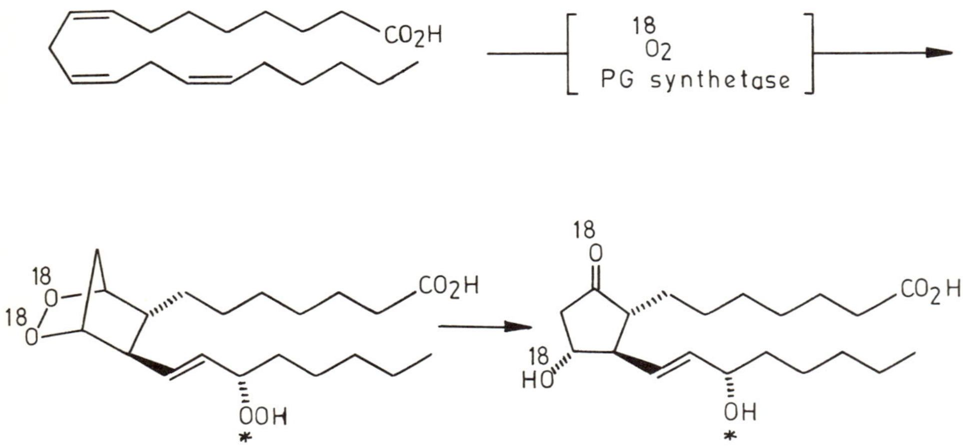

In additional studies, 12-hydroxy, -8,10-heptadecadienoic acid and malonaldehyde were identified as byproducts in the conversion of 8,11,14-eicosatrienoic acid into PGE_1 by microsomes from vesicular gland of sheep.

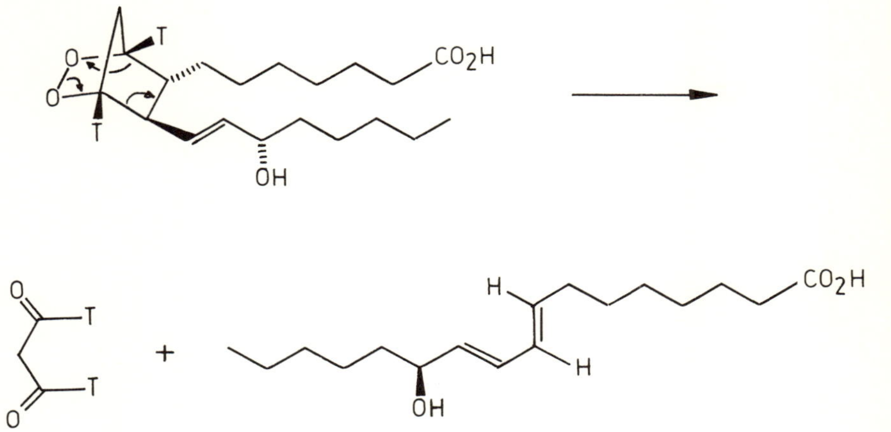

(M. Hamberg and B. Samuelsson, J. Amer. Chem. Soc., 88, 2349 (1966); D. H. Nugteren, R. K. Beerthuis and D. A. van Dorp, Rec. Trav. Chim. Pays-Bas., 85, 405 (1966); M. Hamberg and B. Samuelsson, J. Biol. Chem., 242, 5344 (1967)).

Interestingly, the simplest model of the endoperoxide, arising from cyclopentadiene and singlet oxygen, also undergoes a similar fragmentation *in vitro*:

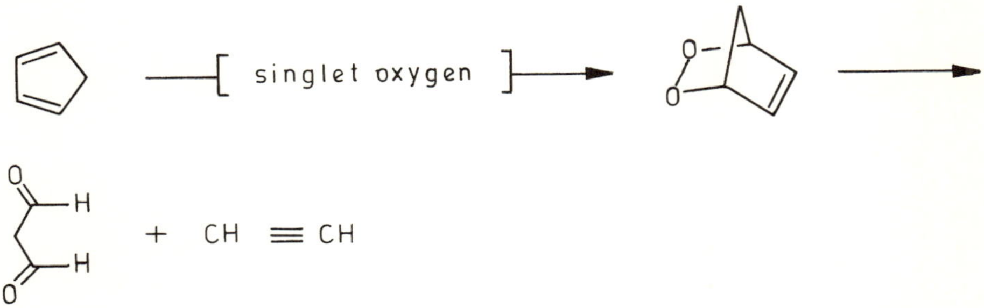

(S. Ranganathan, D. Ranganathan and P. N. Pushpalatha, Unpublished results.)

8. This endoperoxide (PGH_2) was isolated and characterised by its conversion into 11-dehydro $PGF_2\alpha$ and PGE_2 by spontaneous rearrangement on the TLC plate and its conversion into $PGF_2\alpha$ by mild chemical reduction.

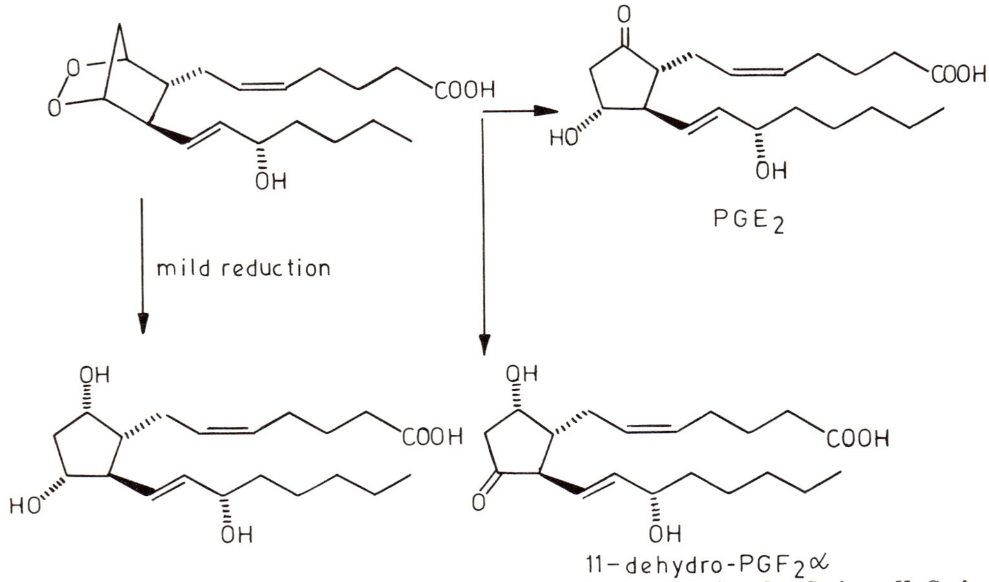

(M. Hamberg and B. Samuelsson, Proc. Nat. Acad. Sci., U.S.A., 70, 899 (1973).

9. Prostaglandin-endoperoxide isomerase catalyses rearrangement of the endoperoxide intermediate into PGE compounds.

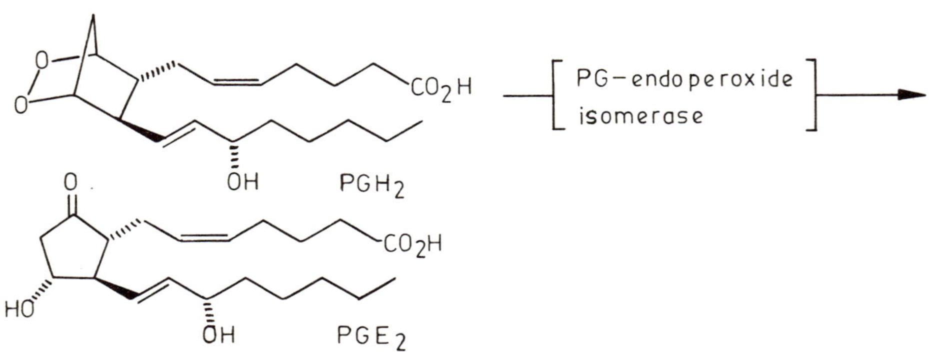

PYRUVIC ACID → ACETYL-CoA

Acetyl-CoA, which is involved-directly or indirectly - in nearly every biosynthetic activity, arises from pyruvic acid by a thiamine pyrophosphate induced α-keto acid decarboxylation for which there is no parallel in organic chemistry.[1]

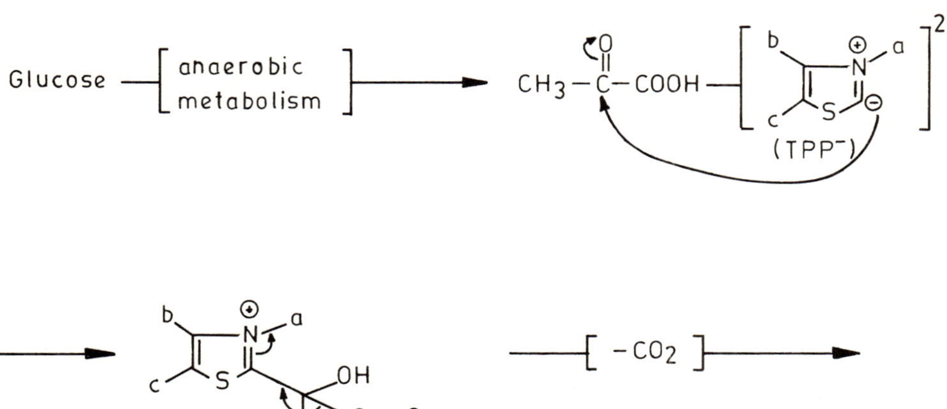

PYRUVIC ACID → ACETYL CoA

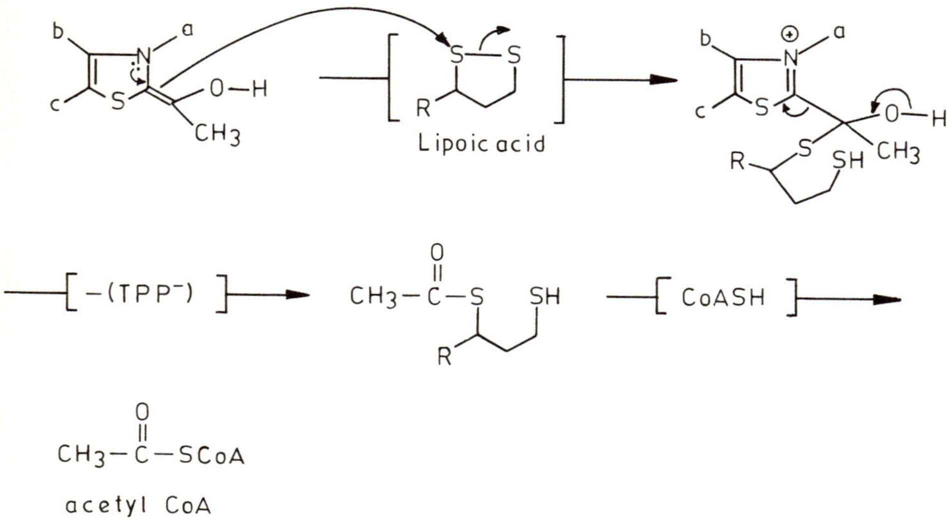

---

1. See, 'Biological Chemistry', Second Edition, Eds. H.R. Mahler and E. H. Cordes, Harper and Row, New York, 1971, p. 519.

2. Thiamine pyrophosphate is involved in many unusual reactions, such as, in the transfer of sugar residues and in the transfer of electrons from pyruvate to ferredoxin, which in turn reduces bound nitrogen.

QUININE

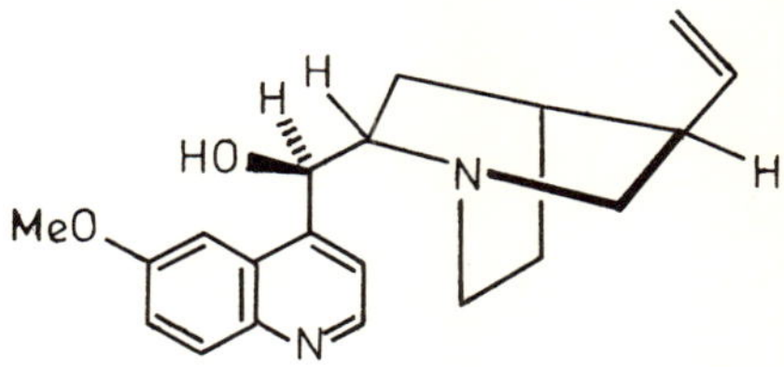

The art of organic synthesis of naturally occurring molecules was initiated with quinine.[1] Only after three decades, when the chemical simulation of nature's art of organic synthesis is receiving increased attention, the biosynthetic scheme of quinine involving the fascinating indole-quinoline transformation of intermediate arising from a secologanin-tryptamine union, was discerned.[2]

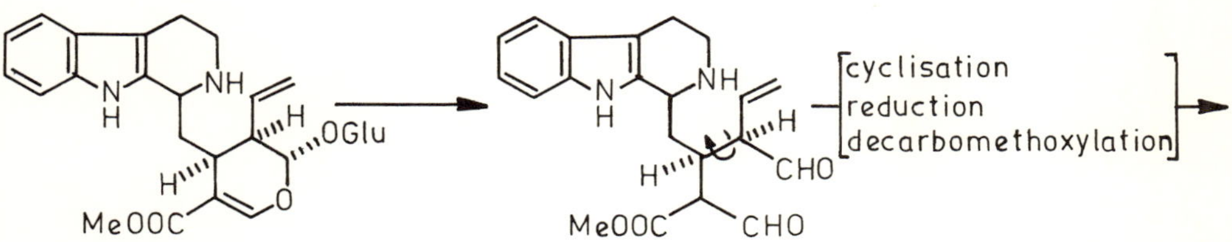

Vincoside (see p. 178)

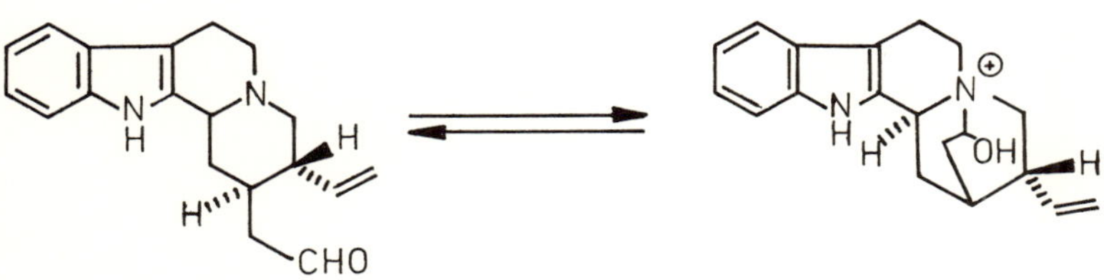

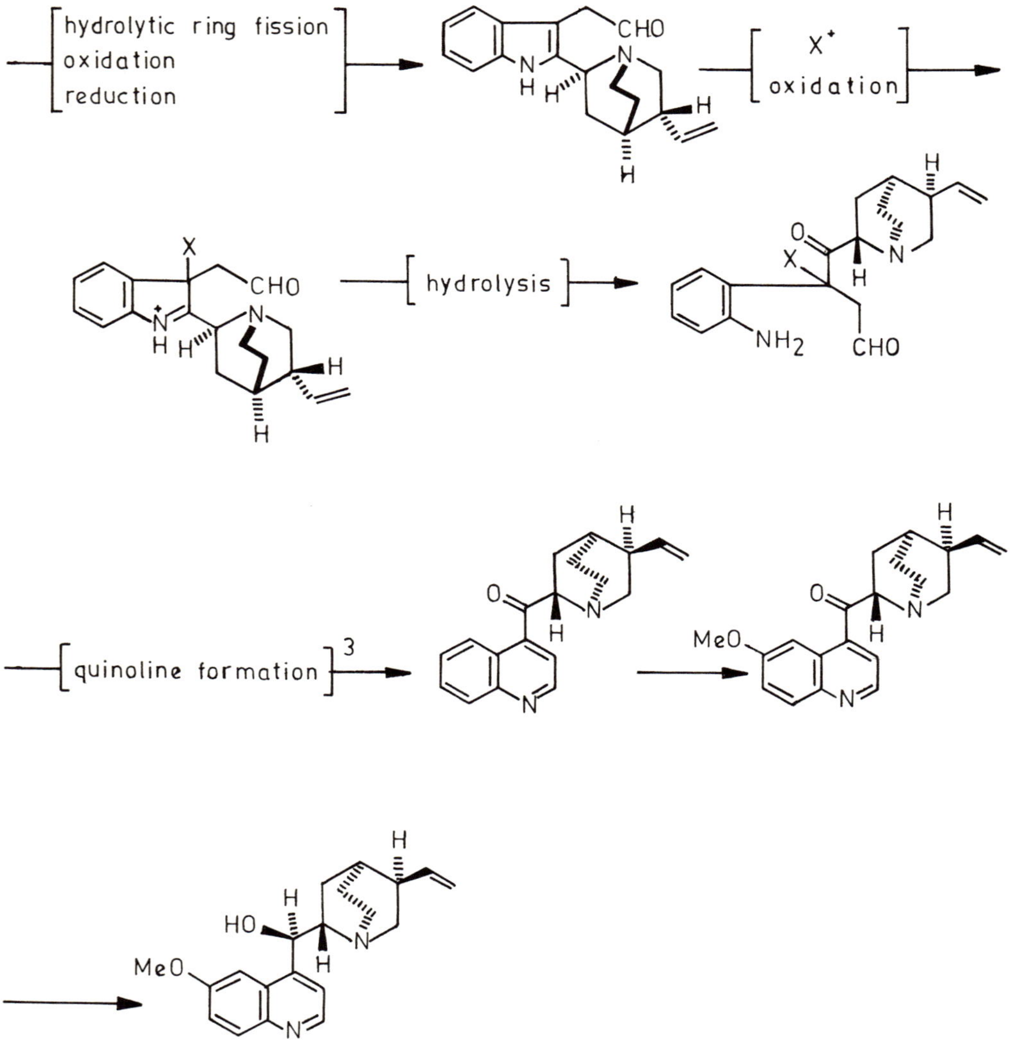
hydrolytic ring fission
oxidation
reduction
CHO
X+
oxidation
CHO
hydrolysis
NH2
CHO
quinoline formation
3
MeO
HO
MeO

Quinine

1. R. B. Woodward and W. E. Doering, J. Amer. Chem. Soc., 66, 849 (1944); ibid., 67, 860 (1945).

2. N. Kowanko and E. Leete, J. Amer. Chem. Soc., 84, 4919 (1962); A. R. Battersby, R. T. Brown, R. S. Kapil, J. A. Knight, J. A. Martin and A. O. Plunkett, Chem. Comm., 888 (1966); E. Leete and J. N. Wemple, J. Amer. Chem. Soc., 91, 2698 (1969); E. Leete, Acc. Chem. Res. 2, 59 (1969); A. R. Battersby and E. S. Hall, Chem. Comm., 194 (1970); A. R. Battersby and R. J. Parry, Chem. Comm., 30, 31 (1971).

3. The indole → quinoline change, initiated with oxidizing agents, has now been recognised as a general biosynthetic as well as synthetic transformation (E. E. van Tamelen and V. B. Haarstad, Tetrahedron Letters, 390 (1961); A. G. Schultz, Chem. Rev., 73, 385 (1973)).

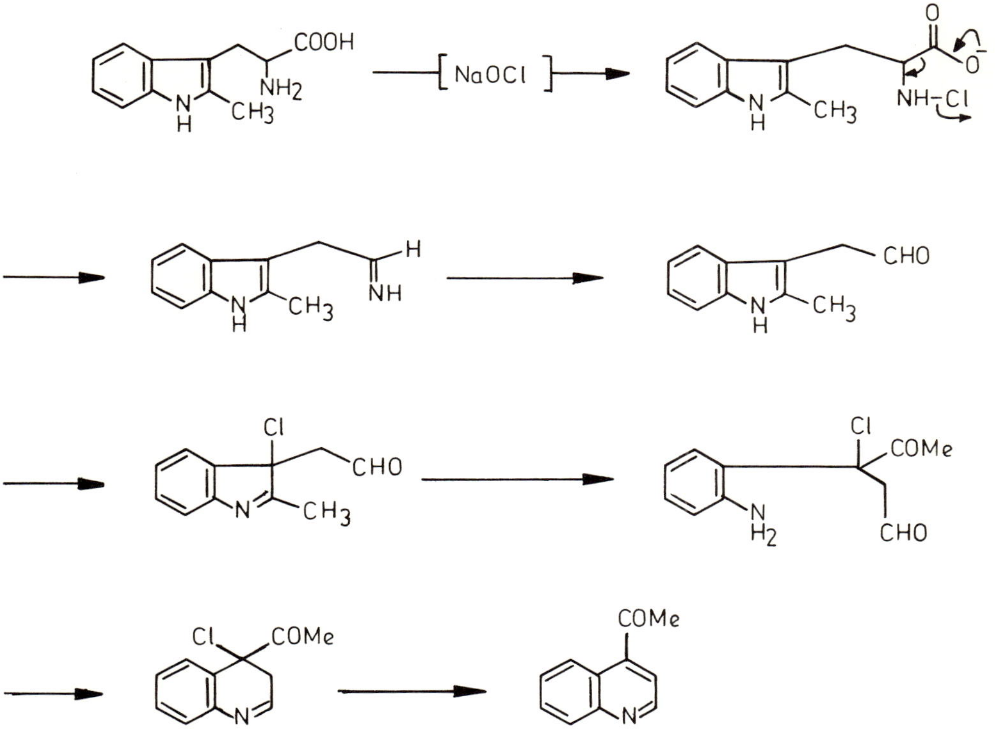

ROTENONE

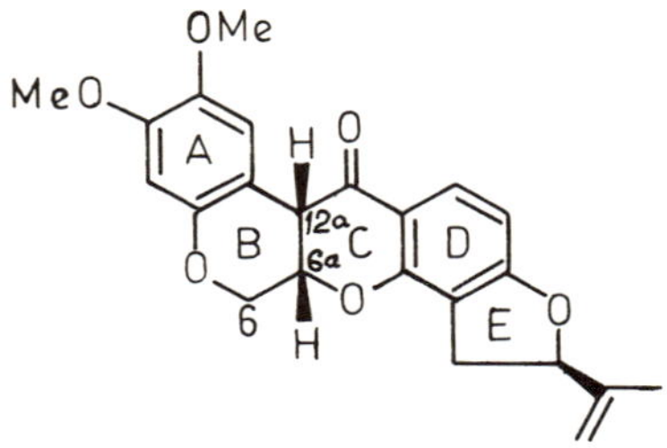

The biosynthesis of pentacyclic Rotenone[1] illustrates the sequential and logical creation of three different types of oxygen heterocycles!

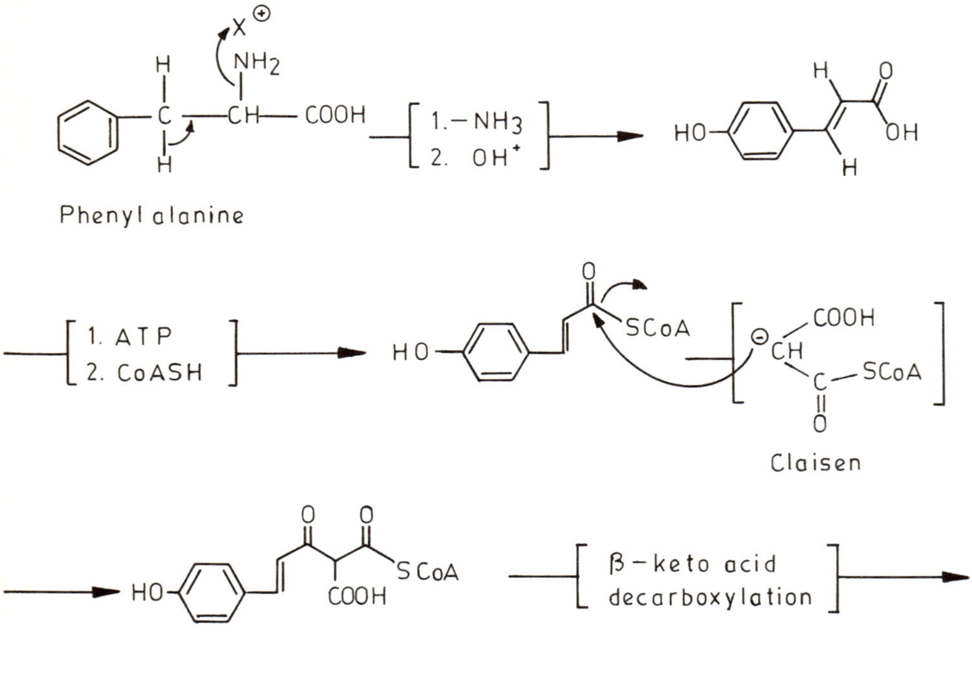

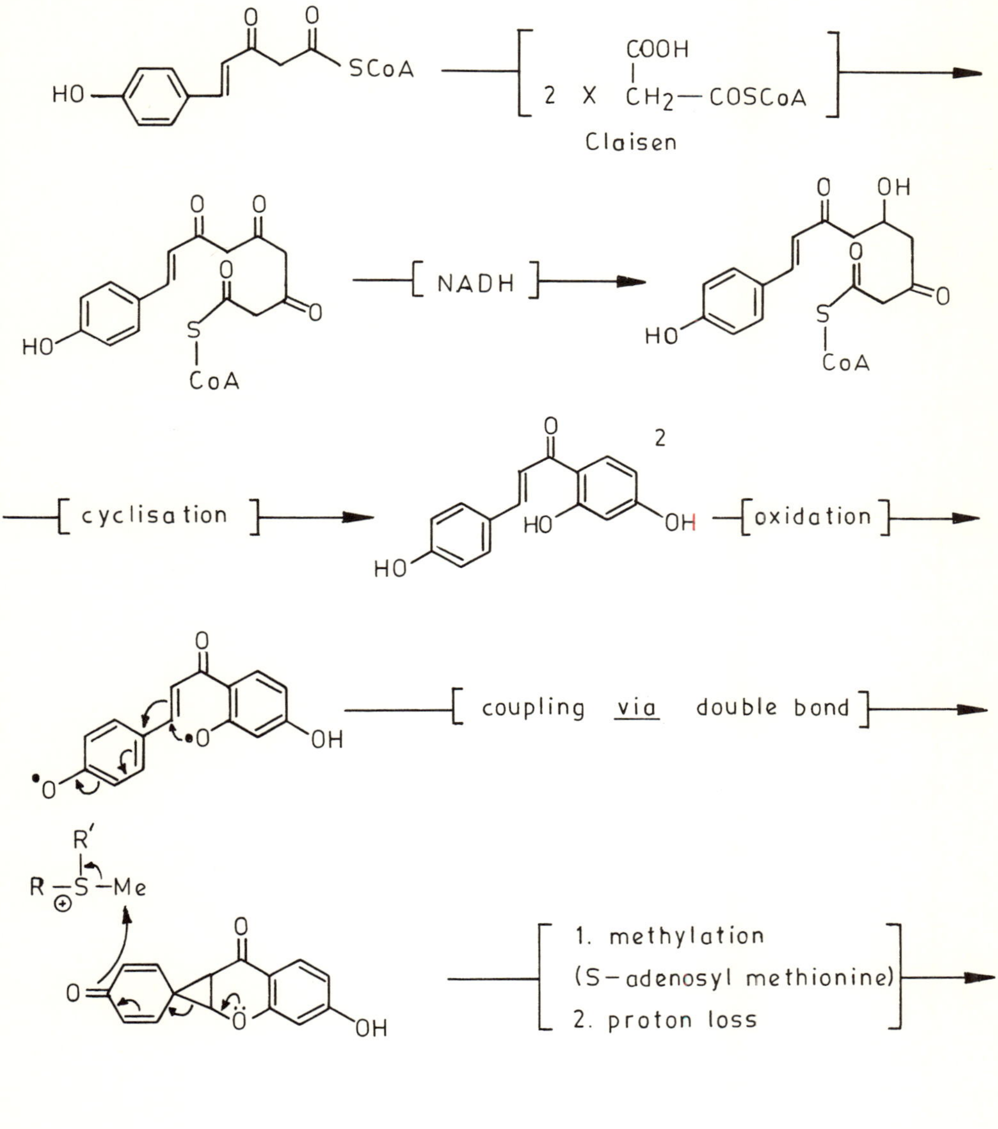
SCoA
HO
2 X
COOH
CH2—COSCoA
Claisen
CoA
NADH
cyclisation
oxidation
2
coupling via double bond
R′
R
S
Me
1. methylation
(S–adenosyl methionine)
2. proton loss
OH

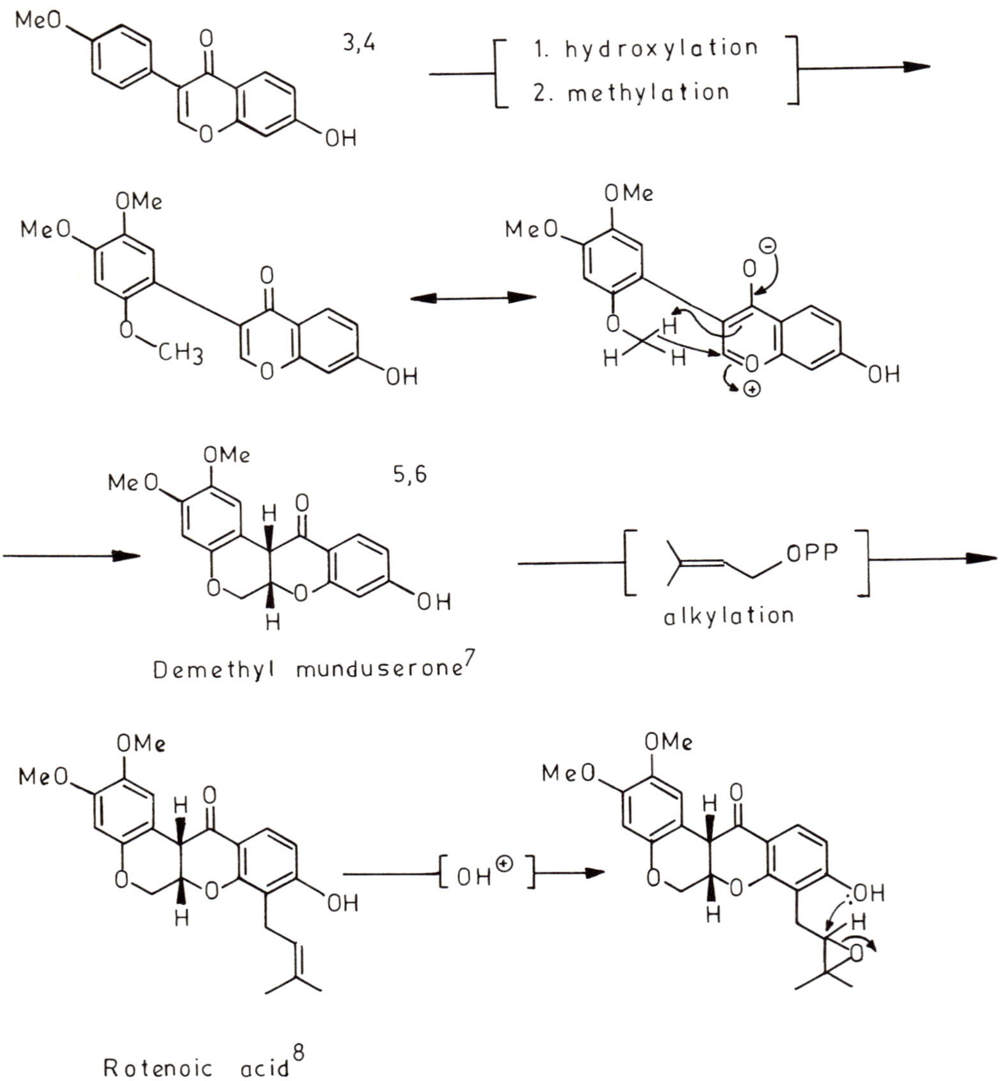
MeO
O
OH
3,4
1. hydroxylation
2. methylation
OMe
CH3
5,6
OPP
alkylation
Demethyl munduserone7
Rotenoic acid8
OH⊕

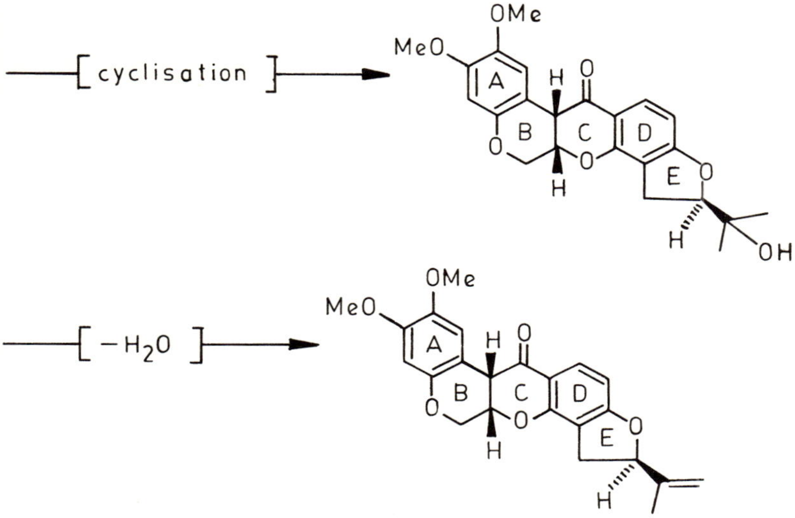

Rotenone

---

1. L. Crombie and M. B. Thomas, J. Chem. Soc. (C), 1796 (1967); L. Crombie, C. L. Green and D. A. Whiting, J. Chem. Soc. (C), 3029 (1968).

2. [^{14}C] labelled 2',4,4'-trihydroxy chalcone was incorporated efficiently into Amorphigenin in A. fruticosa (L. Crombie, P. M. Dewick and D. A. Whiting, Chem. Comm., 1183 (1971)).

3. The involvement of the 4-hydroxyl group in this unusual aryl migration has been confirmed (A. Pelter, Tetrahedron Letters, 897 (1968); A. Pelter, J. Bradshaw and R. F. Warren, Phytochemistry, 10, 835 (1971)). Alternate speculation for this migration are available (K. Freudenberg, G. Carrera and E. Cohn, Annalen, 446, 87 (1926); F. M. Dean, A. McGookin, B. M. Marshall and A. Robertson, J. Chem. Soc., 4573 (1954); H. Grisebach and W. D. Ollis, Experientia, 17, 4 (1961); H. Grisebach and W. Barz, Chem. Ber., 97, 1688 (1964); W. D. Ollis, K.L. Ormand and I. O. Sutherland, J. Chem. Soc. (C), 119 (1970); W. D. Ollis, K. L. Ormand, B. T. Redman, R. J. Roberts and I. O. Sutherland, J. Chem. Soc., (C), 125 (1970)).

4. 7-Hydroxy-4'-methoxy isoflavone is very efficiently incorporated (Ref. 2).

5. This cyclisation can also be rationalised on the basis of radical intermediates (See, p. 81, footnote 3).

6. The methoxyl carbon has been unequivocally linked to the methylene group of the newly formed ring:

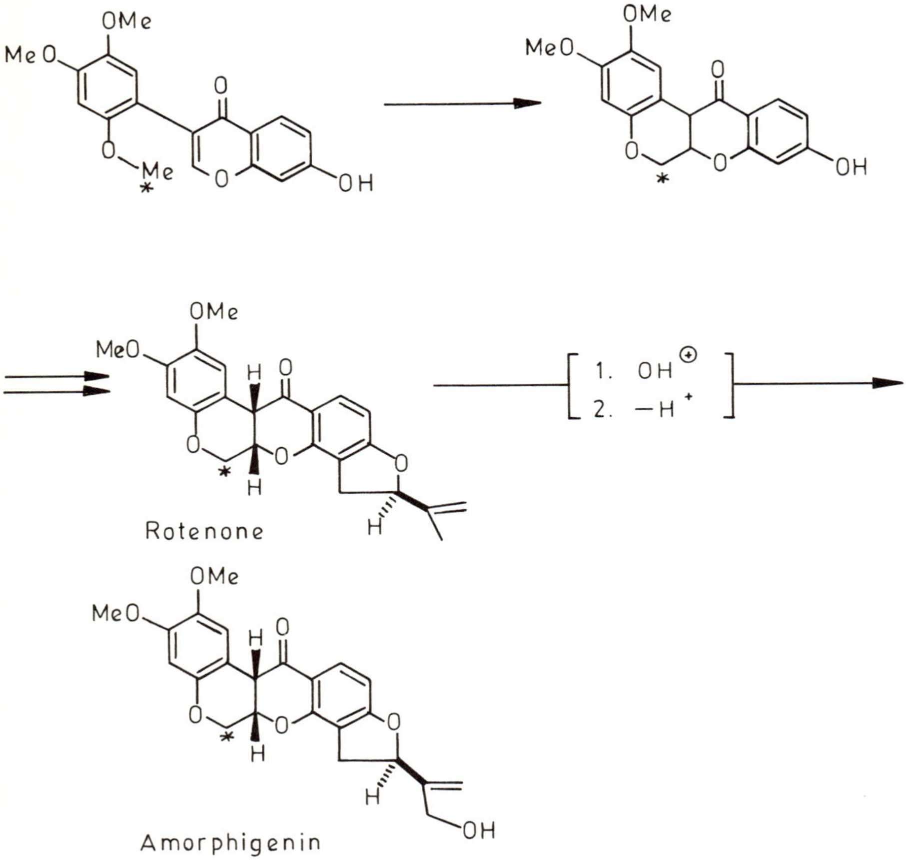

Rotenone

Amorphigenin

(L. Crombie, P. M. Dewick and D. A. Whiting, Chem. Comm., 1469 (1970).)

7. (±)-[6-^{3}H]-9-Demethylmunduserone is a good precursor for amorphigenin in *A. fruticosa* (Ref. 2).

8. (±)-[6-^{3}H] rotenoic acid is converted into amorphigenin (Ref. 2).

SPARTEINE

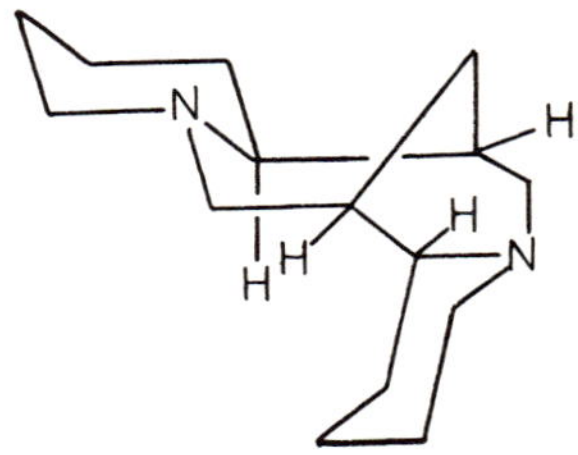

The biosynthesis of Lycopodine and Sparteine from lysine highlights two aspects relating to the creation of highly condensed nitrogen heterocycles from simple precursors. Of these the biosynthesis of Sparteine is prettier since this nitrogen base is exclusively built from lysine![1]

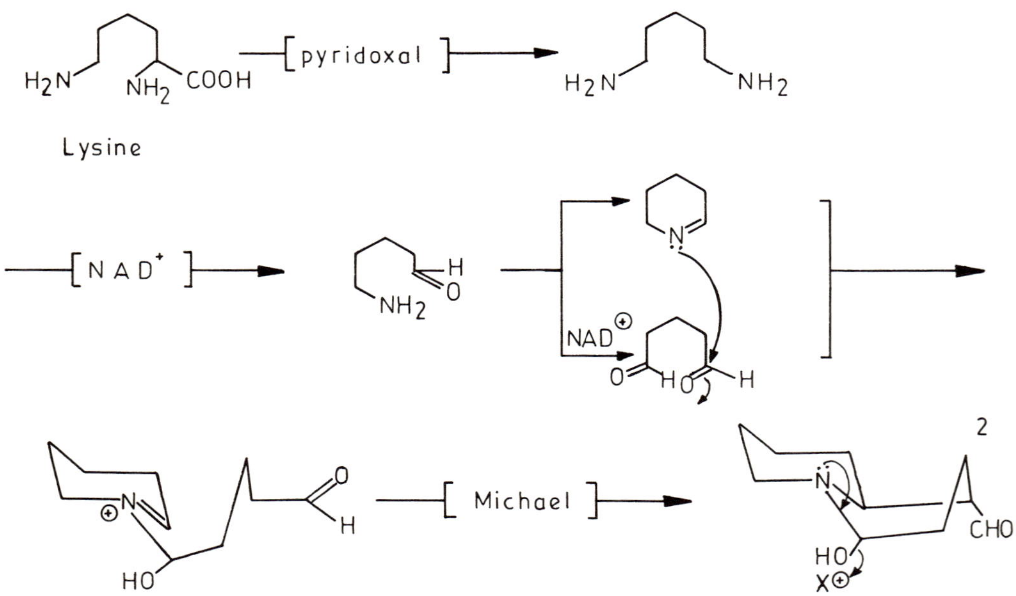

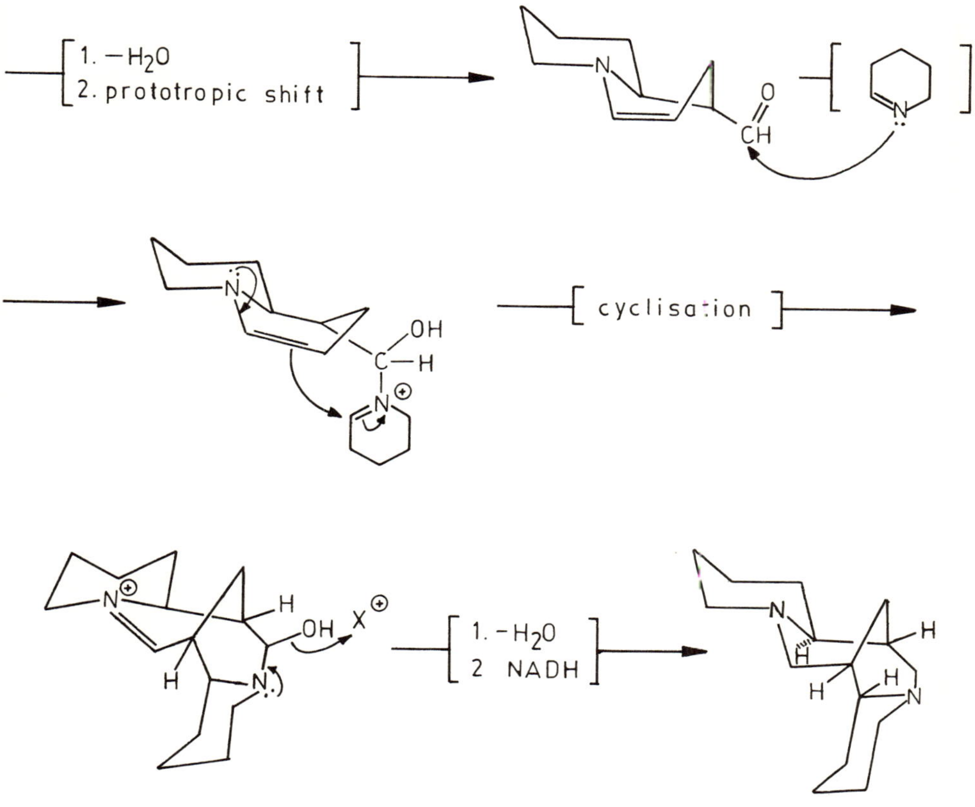

Sparteine

1. See, D. W. Hughes and K. Genest in 'Phytochemistry', Vol.II, Ed. L. P. Miller, Van Nostrand Reinhold, New York, 1973, p. 128 and references cited therein.

2. Reduction of this intermediate will yield Lupenine :

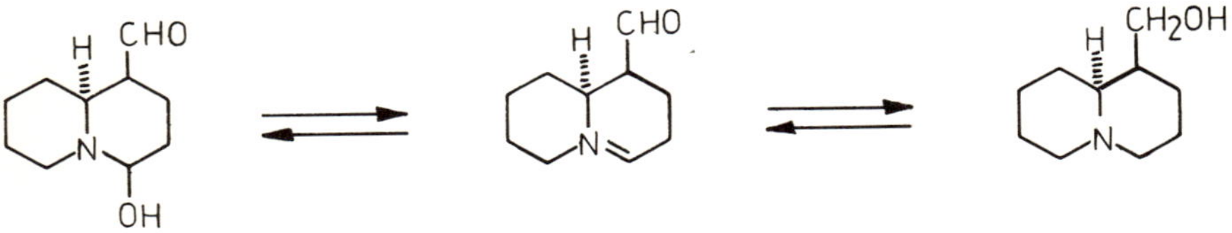

Lupenine

Lupenine is a biosynthetic precursor to sparteine, thus implying the operation of a reverse oxidation sequence.

SQUALENE

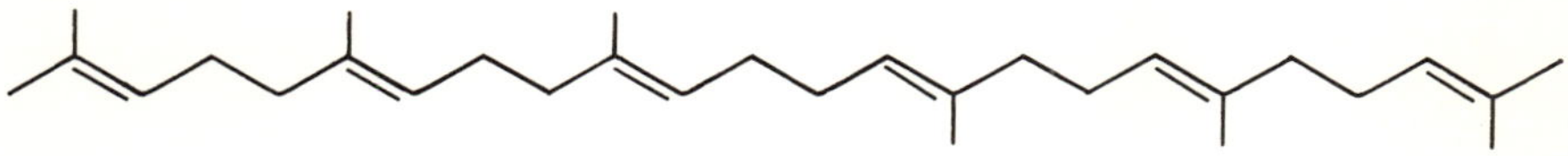

Nature's strategy[1] relating to the creation of squalene illustrates both the head-tail and the tail-tail unions that operate in the construction of poly-isoprenoids from the isoprene unit that is latent in Mevalonic acid:

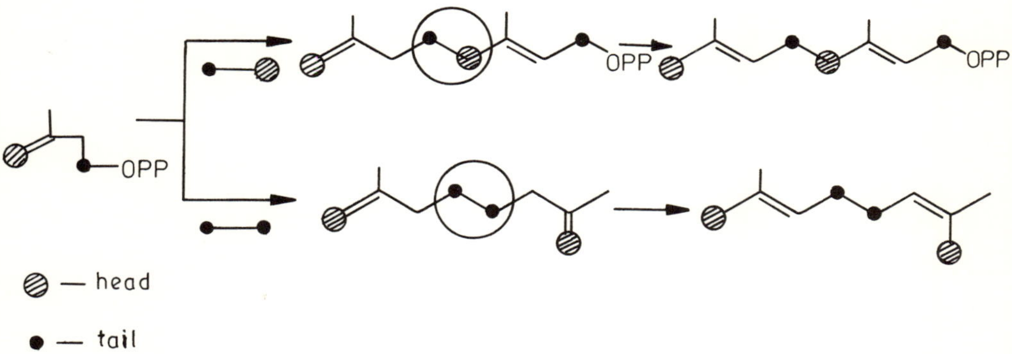

Whilst the head-tail union can be readily understood on the basis of π participation (vide infra), the tail-tail union illustrates a magnificent facet of the art in biosynthesis via a double cyclopropane rearrangement!

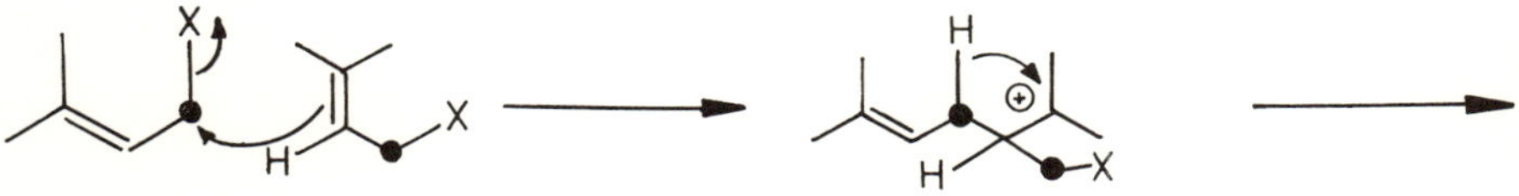

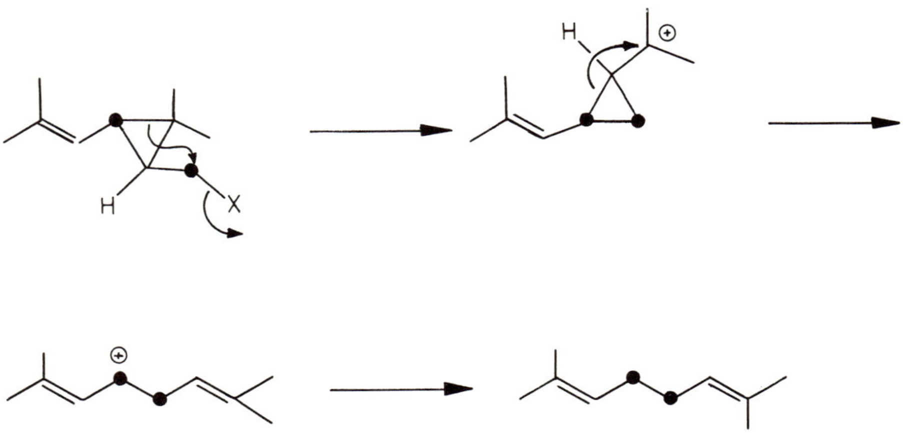

Yet another fascinating feature of the tail-tail union in Squalene is that the enzyme system can recognise the prochiral protons of the tail and the coupling process is attended with retention of the stereochemistry of one unit whilst bringing about a highly stereospecific inversion of the other.[2]

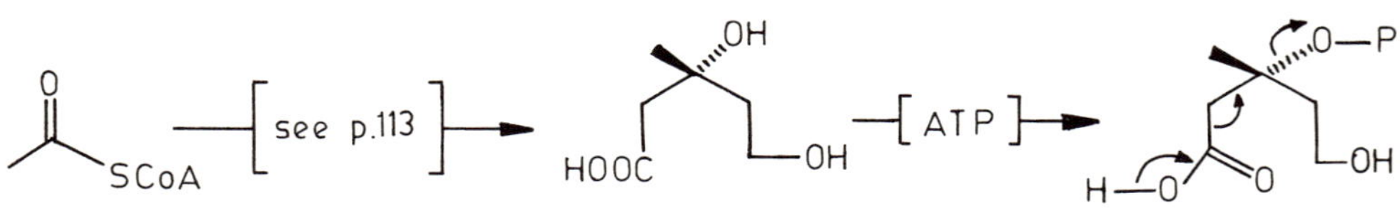

Acetyl - coenzyme A

Mevalonic acid

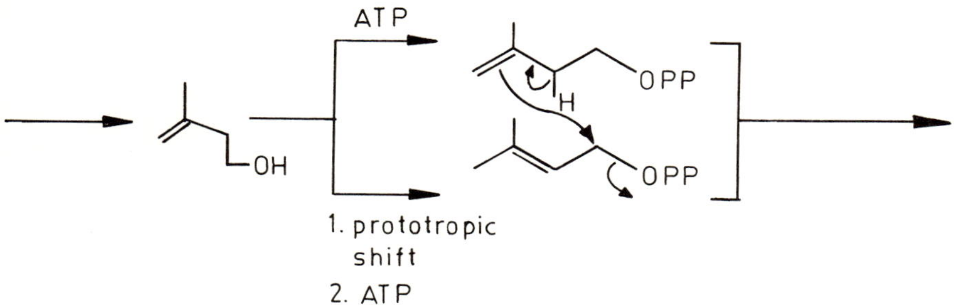

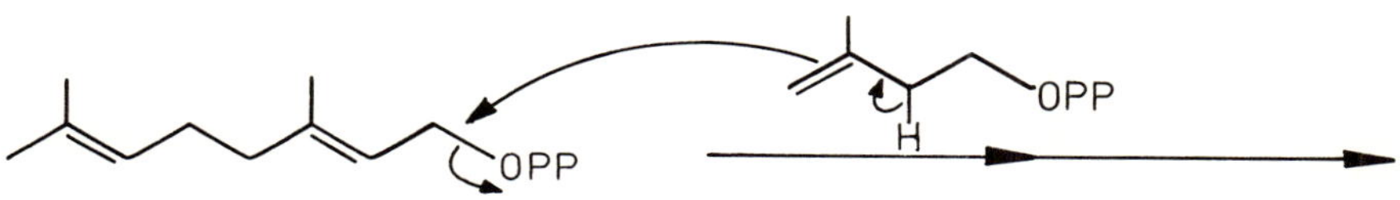

Geranyl pyrophosphate

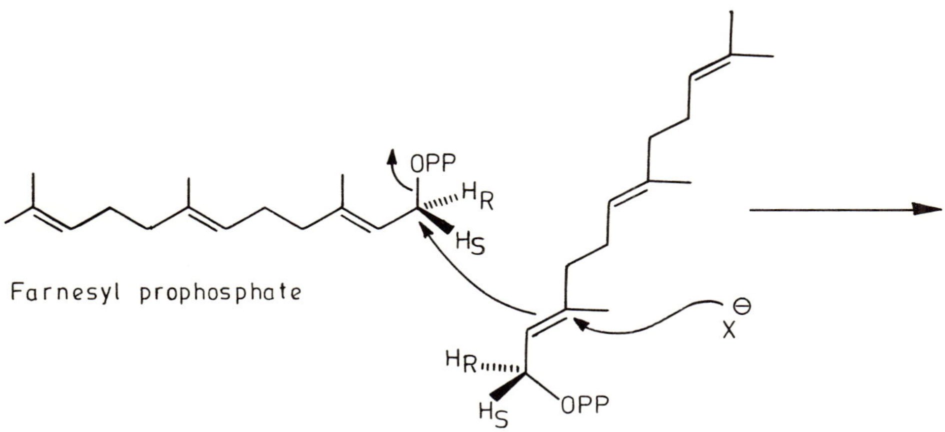

Farnesyl prophosphate

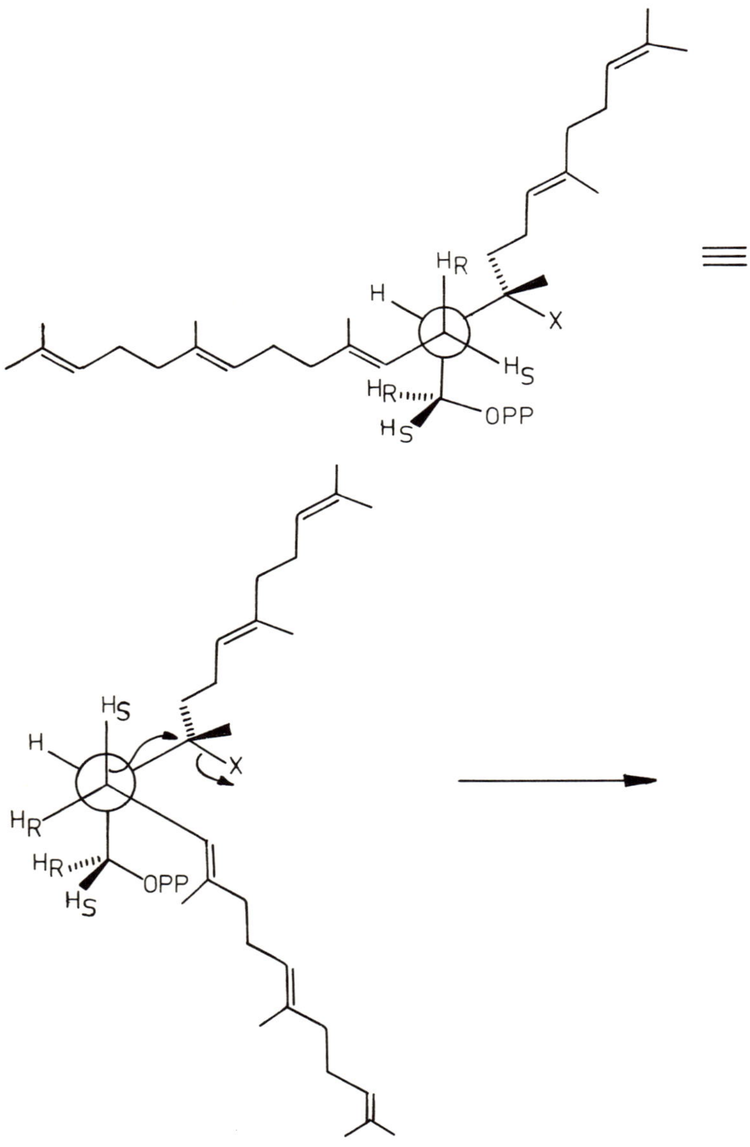
HR
H
X
HS
HR
OPP
HS
HS
H
X
HR
HR
OPP
HS

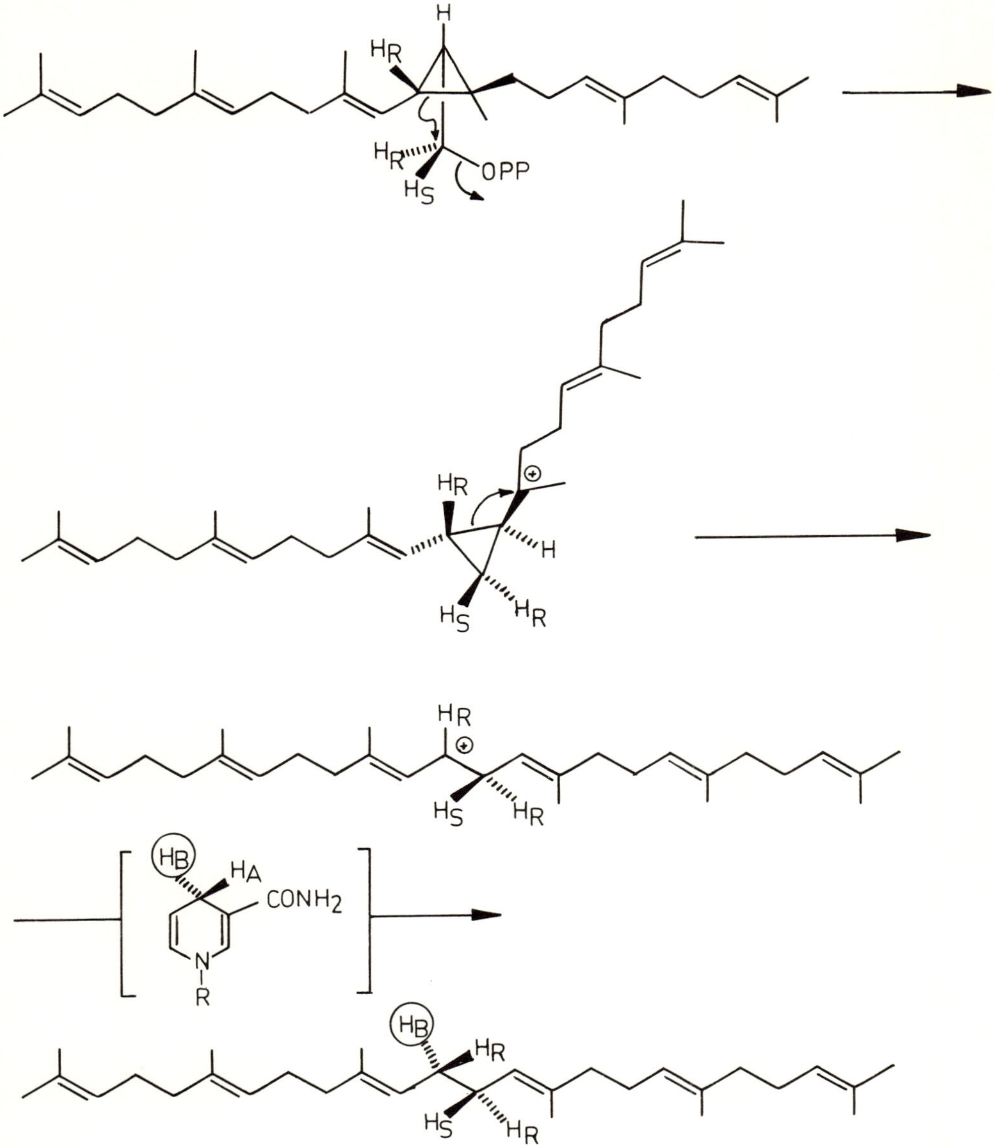

Squalene

1. R. B. Clayton, Quart. Rev., 19, 168 (1965) and references therein; J. W. Cornforth, Quart. Rev., 23, 125 (1969); L. J. Mulheirn and P. J. Ramm, Chemical Society Reviews, 1, 259 (1972).

2. W. W. Epstein and H. C. Rilling, J. Biol. Chem., 245, 4597 (1970); R. V. M. Campbell, L. Crombie and G. Pattenden, Chem. Comm., 218 (1971); L. J. Altman, R. C. Kowerski and H. C. Rilling, J. Amer. Chem. Soc., 93, 1782 (1971); H. C. Rilling, C. D. Poulter, W. W. Epstein and B. Larsen, ibid., p. 1783; R. M. Coates and W. H. Robinson, ibid., p. 1785.

STERIGMATOCYSTIN, AFLATOXIN-B_1

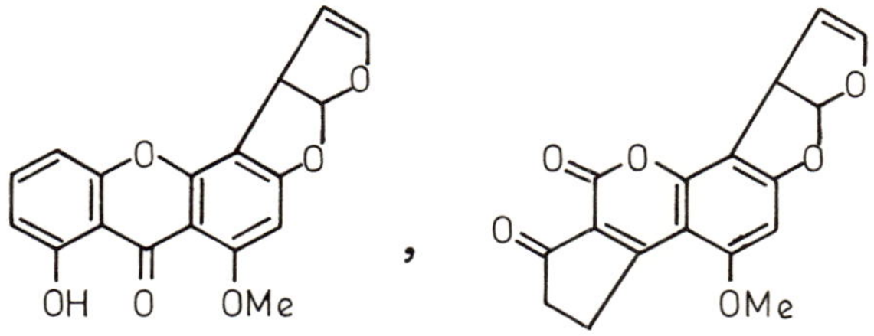

Sterigmatocystin , Aflatoxin B_1

The biosynthesis of Sterigmatocystin[1] and its further elaboration to the carcinogenic mold metabolite Aflatoxin B_1[2] highlights the fascinating, intricate, molecular rearrangements that are often associated with phenolic and polyphenolic substrates.

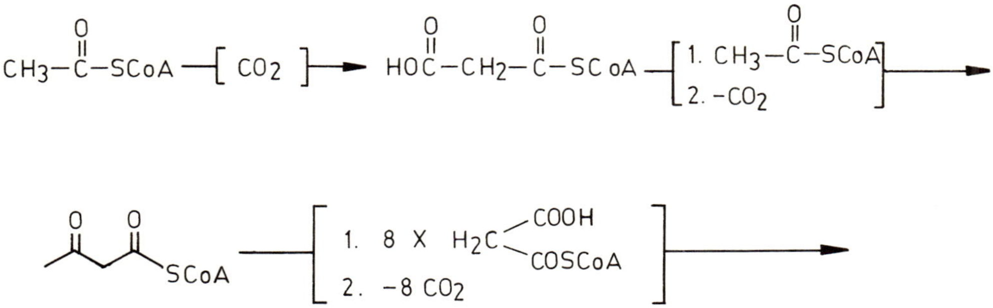

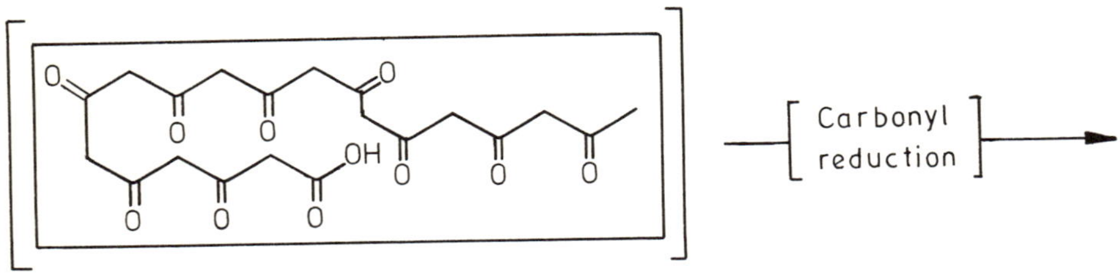

matrix bound C_{20}-polyketide

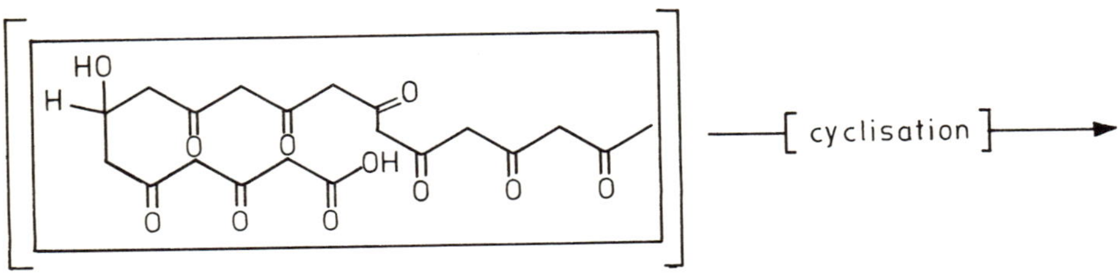

matrix bound C_{20}-polyketide

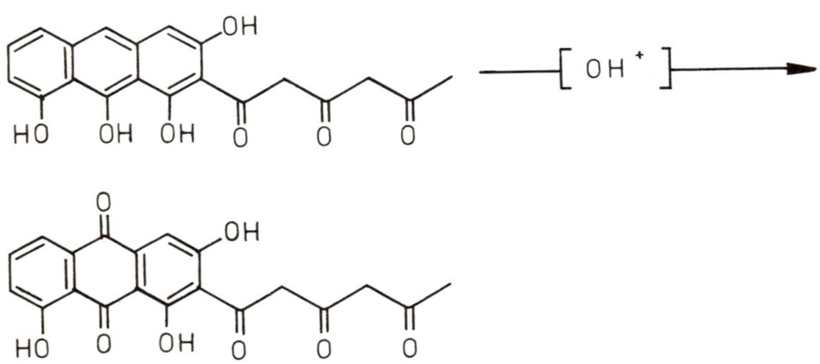

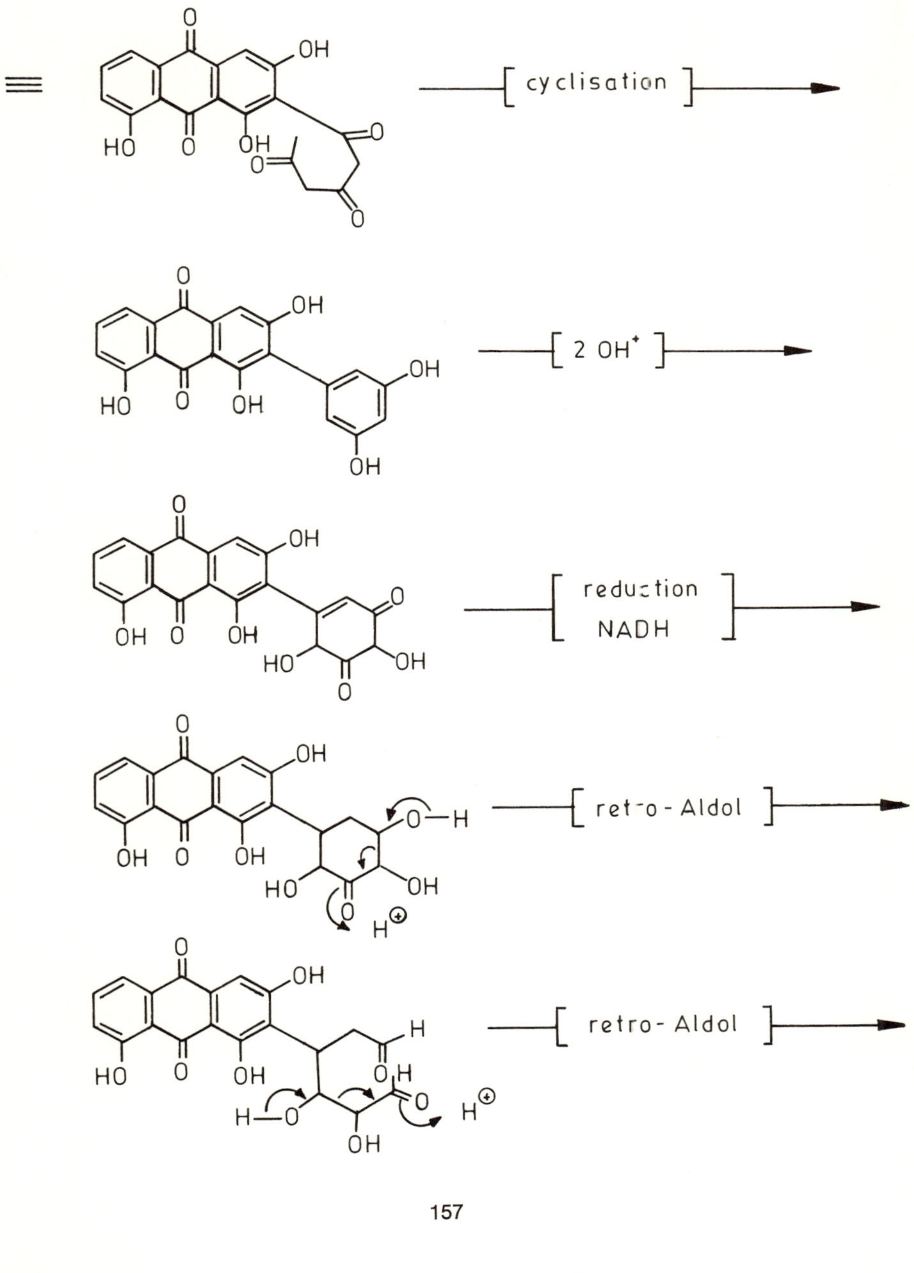
cyclisation
2 OH$^+$
reduction
NADH
ret-o-Aldol
retro-Aldol

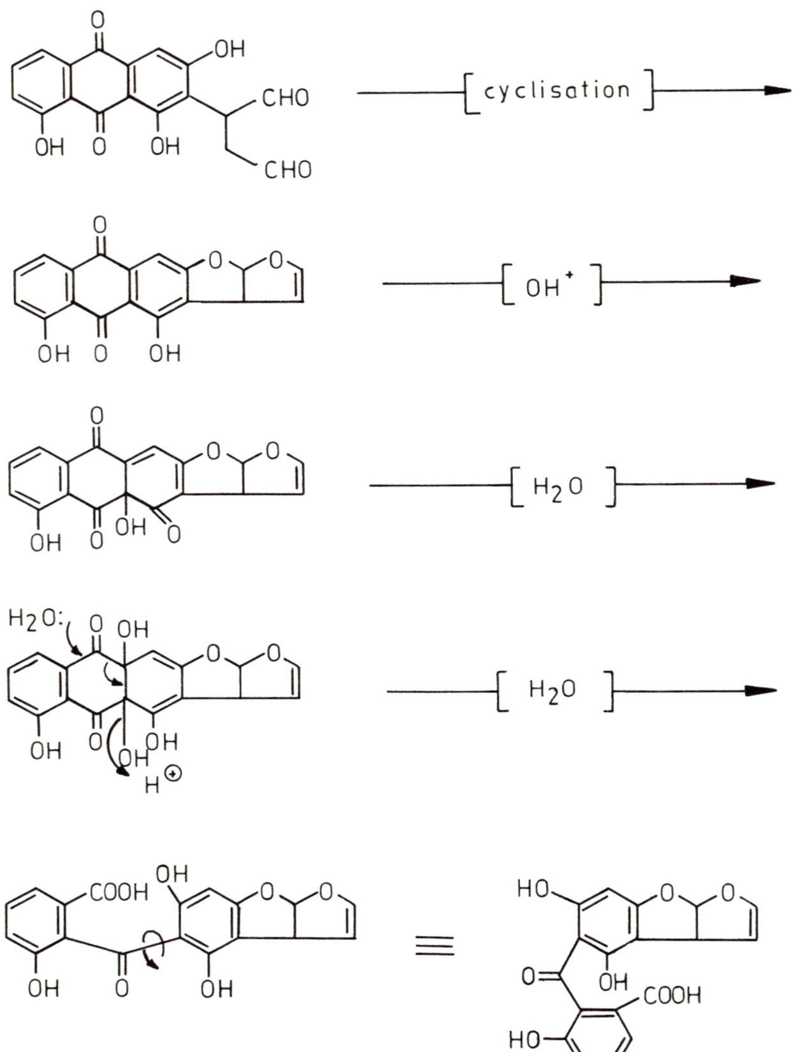

O
OH
CHO
OH O OH
CHO
[cyclisation]
O
O O
OH O OH
[OH⁺]
O
O O
OH O OH O
[H₂O]
H₂O:
O OH
O O
OH O OH OH
H⊕
[H₂O]
OH
COOH
O O
OH O OH
HO
O O
O
OH
COOH
HO

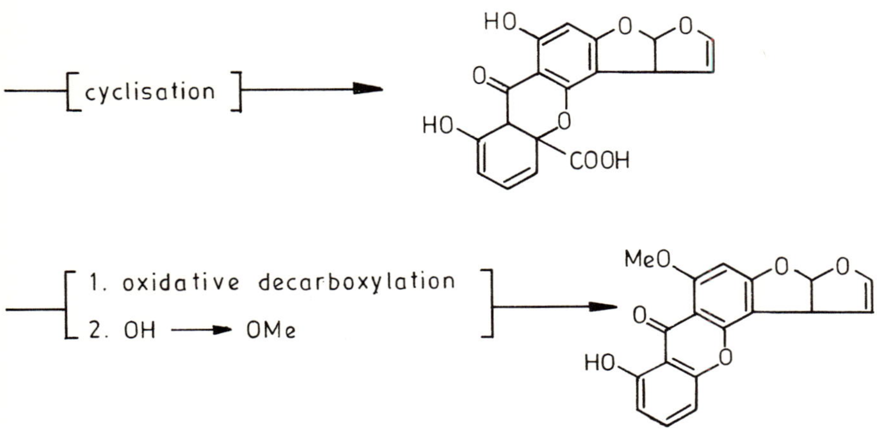

Sterigmatocystin

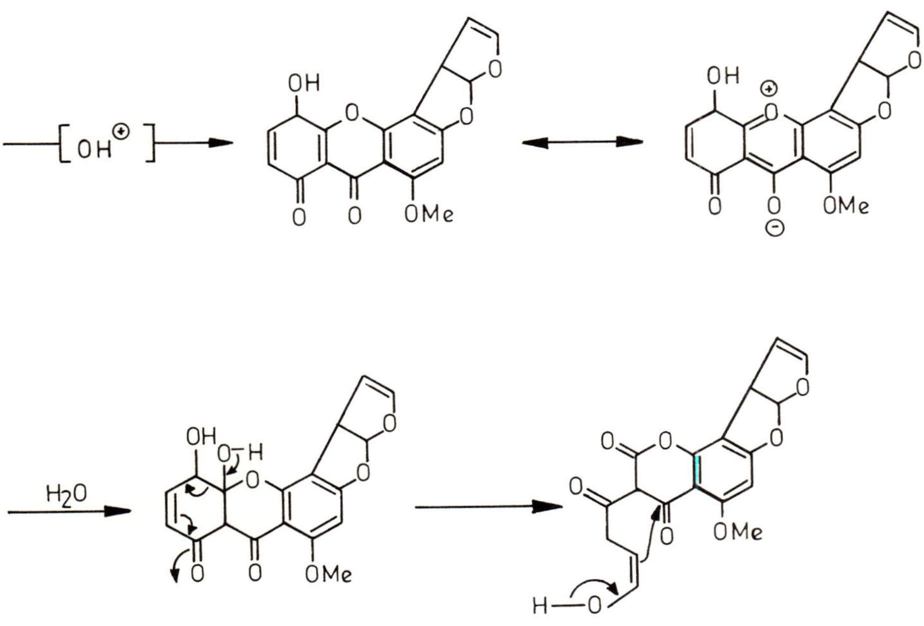

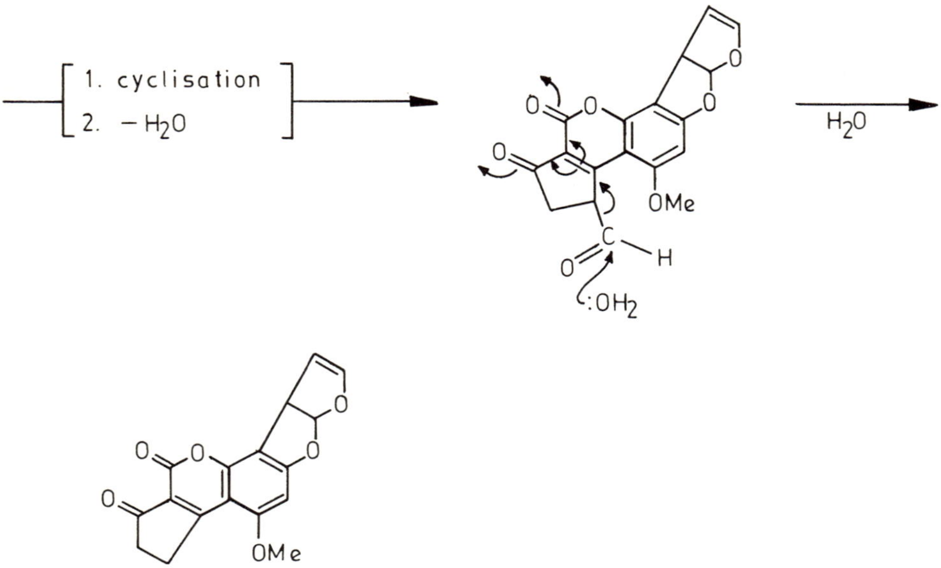

Aflatoxin B$_1$

---

1. P. S. Steyn, R. Vleggaar, P. L. Wessels and D. B. Scott, Chem. Comm., 193 (1975); K. G. R. Pachler, P. S. Steyn, R. Vleggaar and P. L. Wessels, Chem. Comm., 355 (1975).

2. Sterigmatocystin in converted into Aflatoxin-B$_1$ by cultures of A. parasiticus (D. P. H. Hsieh, M. T. Lin and R. C. Yao, Biochem. Biophys. Res. Comm., 52, 992 (1973)).

STIPITATIC ACID

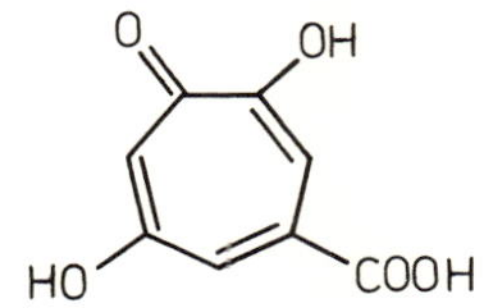

One would have intuitively felt that the art of the construction of the tropolone unit in colchicine must be, in view of the other features already present, special. This is indeed the case as is illustrated with the biosynthesis of a relatively unencumbered tropolone unit that is present in stipitatic acid and confirming[1] the early prediction of Robinson[2] that tropolones arise by oxidative ring expansion of benzenoid precursors.

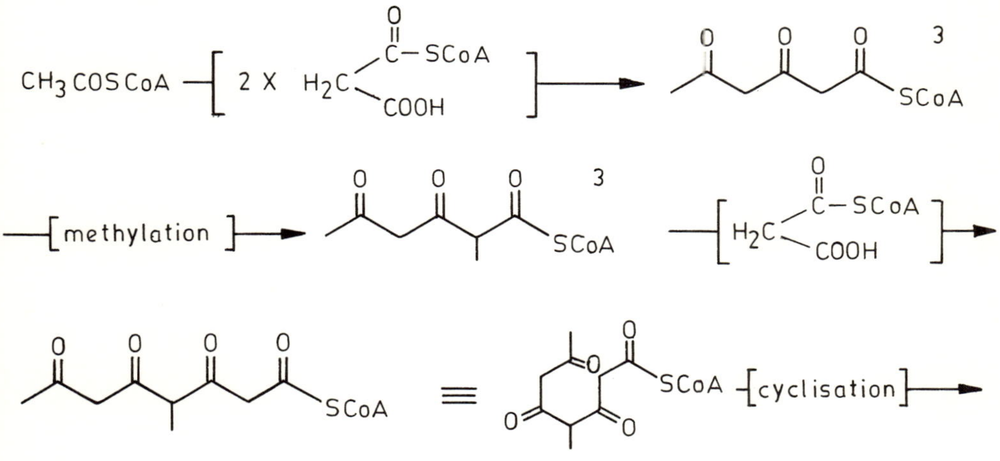

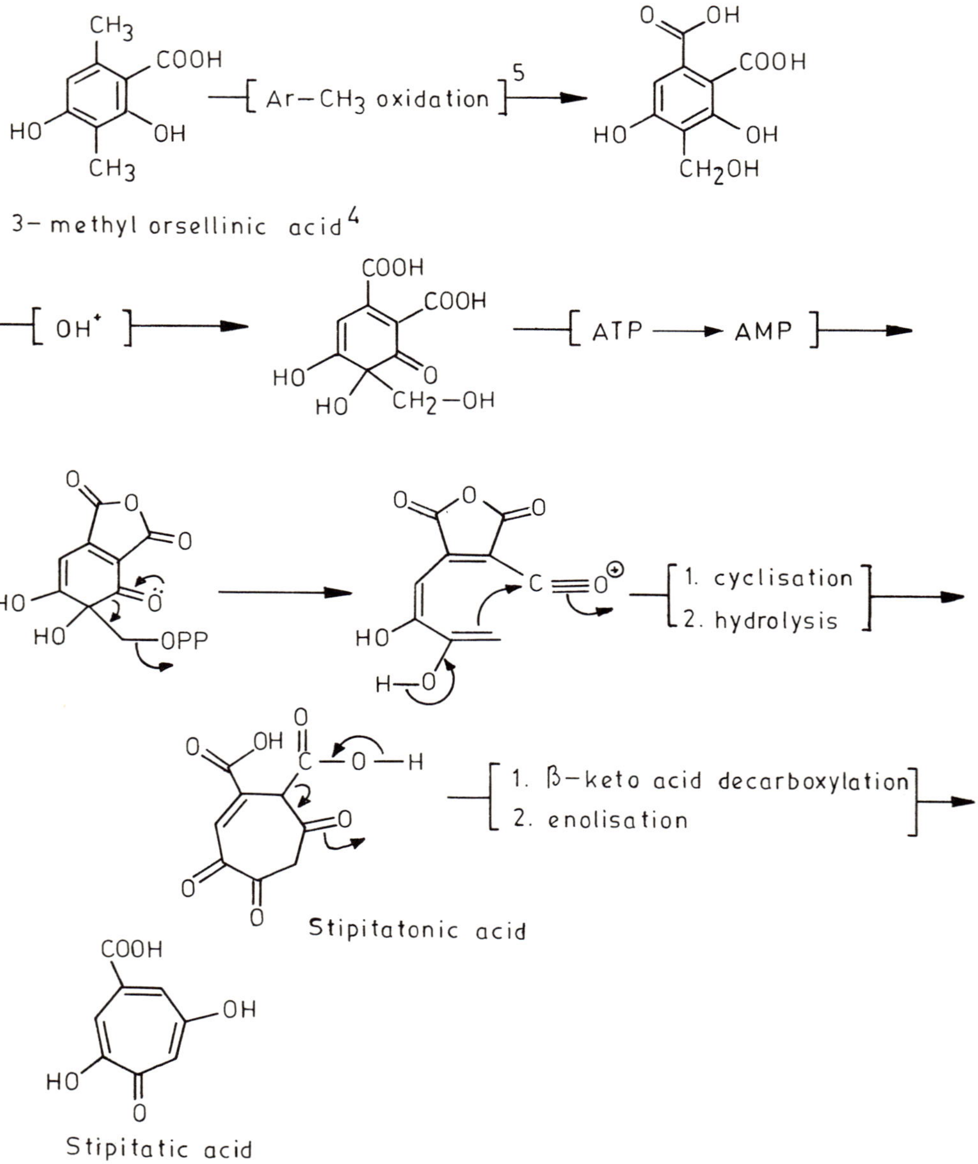
Ar–CH3 oxidation 5
3-methyl orsellinic acid 4
OH+
ATP → AMP
1. cyclisation
2. hydrolysis
1. β-keto acid decarboxylation
2. enolisation
Stipitatonic acid
Stipitatic acid

1. A. I. Scott, H. Guilford and E. Lee, J. Amer. Chem. Soc., 93, 3534 (1971); A. I. Scott and K. J. Wiesner, Chem. Comm., 1075 (1972).

2. R. Robinson, Chem. Ind. (London), 12 (1951).

3. The intermediacy of these polyketide intermediates was suggested by the isolation of their cyclised products during inihibition of tropolone biosynthesis (T. E. Acker, P. E. Brenneisen and S. W. Tanenbaum, J. Amer. Chem. Soc., 88, 834 (1966); R. Bentley and P. M. Switkowits, J. Amer. Chem. Soc., 89, 676 (1967)).

4. Labelled 3-methyl orsellinic acid was efficiently incorporated in stipitatonic and stipitatic acids in *P. stipitatum* (Ref. 1).

5. Monooxygenase mechanism was suggested for the oxidative transformation of the benzenoid to troponoid system based on $^{18}O_2$ and $CH_3{}^{14}C^{18}O_2H$ incorporation experiments (Ref. 1).

STRYCHNINE

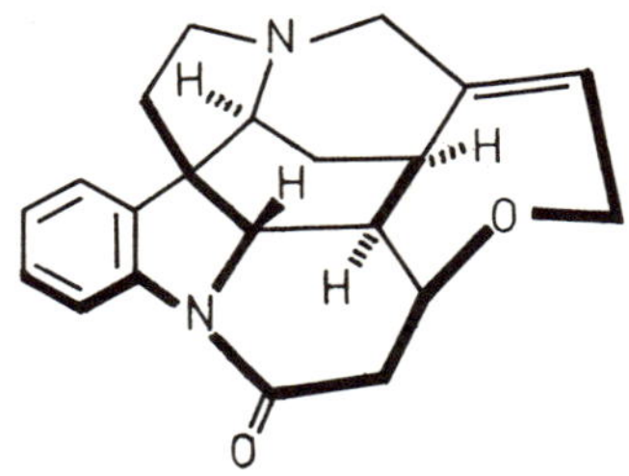

For its size, strychnine has the most complicated framework.[1] It is instructive and amusing to compare the intuitive and daring biogenetic theory put forward by Woodward[2] - along which the strategy for strychnine synthesis was laid out[3] - with the nature's strategy that starts with the highly versatile vincoside.[4]

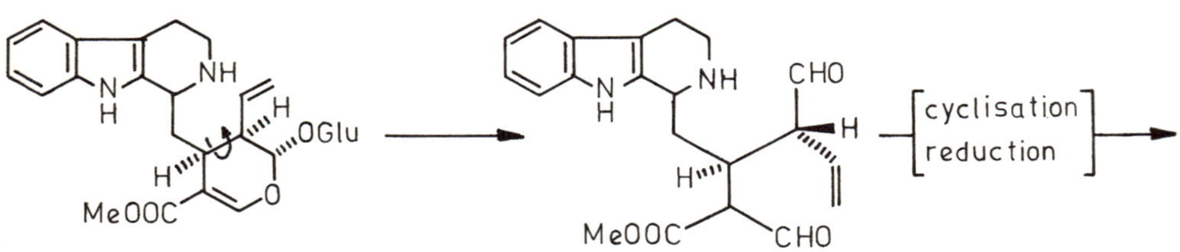

Vincoside (see p.178)

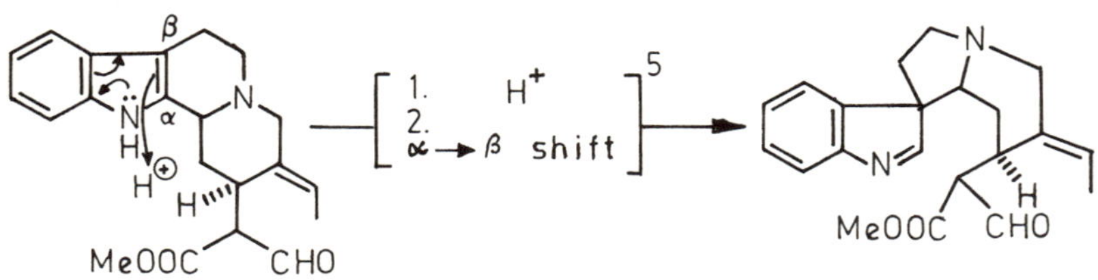

Geissoschizine

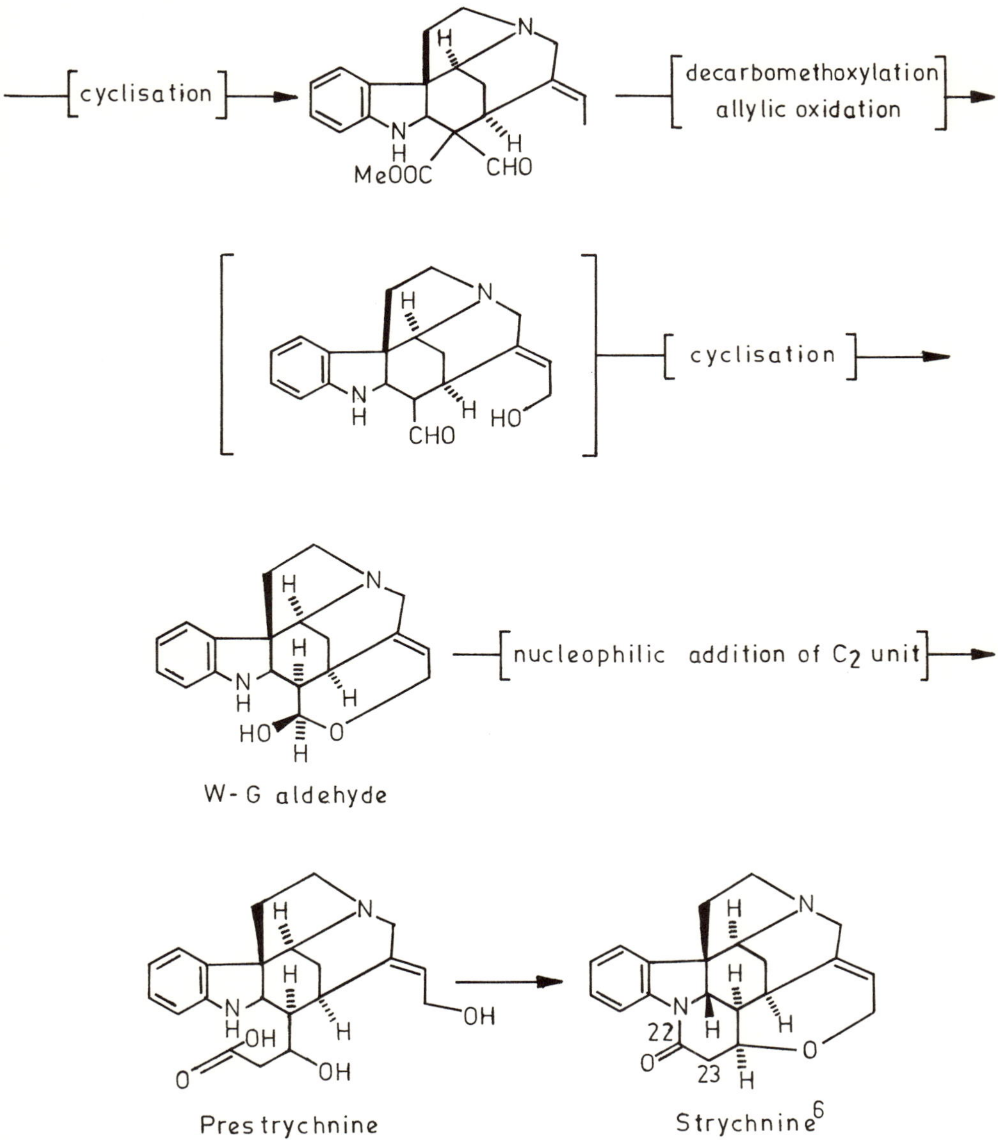
cyclisation
N
H
H
MeOOC
CHO
H
decarbomethoxylation
allylic oxidation
N
H
N
H
H
CHO
HO
cyclisation
N
H
N
H
H
H
HO
O
H
W-G aldehyde
nucleophilic addition of C2 unit
N
H
N
H
H
OH
O
OH
H
OH
Prestrychnine
N
H
N
H
H
H
22
O
23
H
O
Strychnine6

1. R. Robinson in 'Progress in Organic Chemistry', edited by J. W. Cook, Vol. 1, p. 2, Butterworths, London (1952).

2. R. B. Woodward, Nature, 162, 155 (1948).

3. R. B. Woodward, M. P. Cava, W. D. Ollis, A. H. Hunger, H. U. Daeniker and K. Schenker, J. Amer. Chem. Soc., 76, 4749 (1954); Tetrahedron, 19, 247 (1963).

4. ch. Schlatter, E. E. Waldner, H. Schmid, W. Maier and D. Groger, Helv. Chim. Acta, 52, 776 (1969); S. I. Heimberger and A. I. Scott, Chem. Comm., 217 (1973).

5. This pathway involving α protonation followed by $\alpha \rightarrow \beta$ shift has *in vitro* analogy (J. Harley-Mason and W. Waterfield, Tetrahedron, 19, 65 (1963); A. J. Gaskelle and J. A. Joule, *ibid*., 23, 4053 (1967)).

6. The well established W-G aldehyde → prestrychnine → strychnine change is particularly noteworthy since an N-acylated W-G aldehyde (Diaboline) would have normally been considered as a more probable intermediate than prestrychnine (ref. 4).

TETRACYCLINE

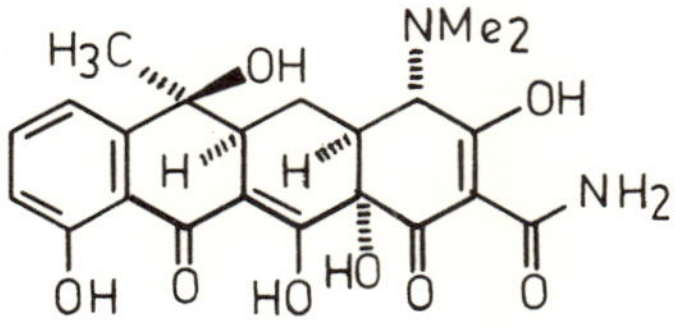

The hydronaphthacenic carbon skeleton of the tetracyclines is built up from acetate in Nature.[1] The malonamoyl-CoA serves as an initiator for a matrix polymerisation of eight molecules of malonyl-CoA to produce a bound linear polyketide-amide, which is then elaborated to tetracyclines by a stepwise, logical and simple sequence.

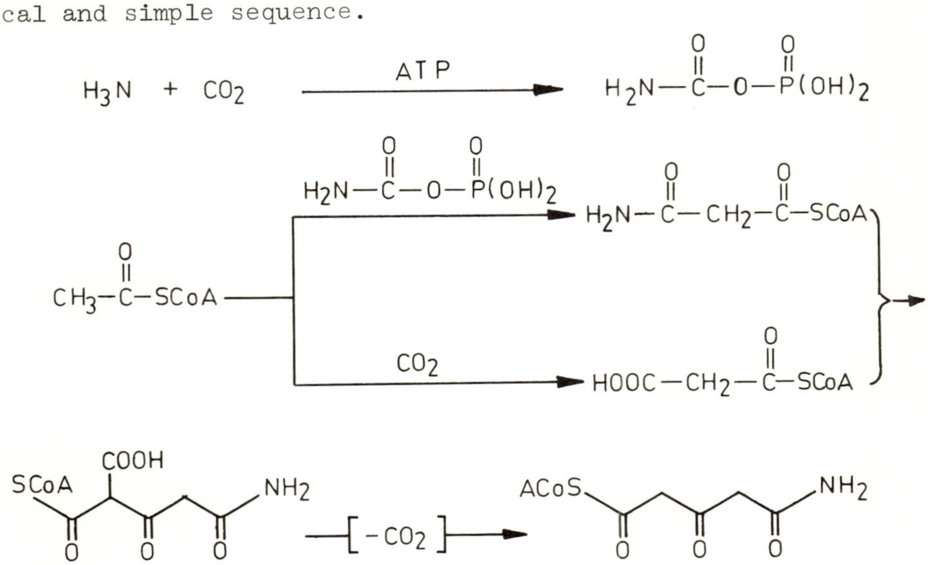

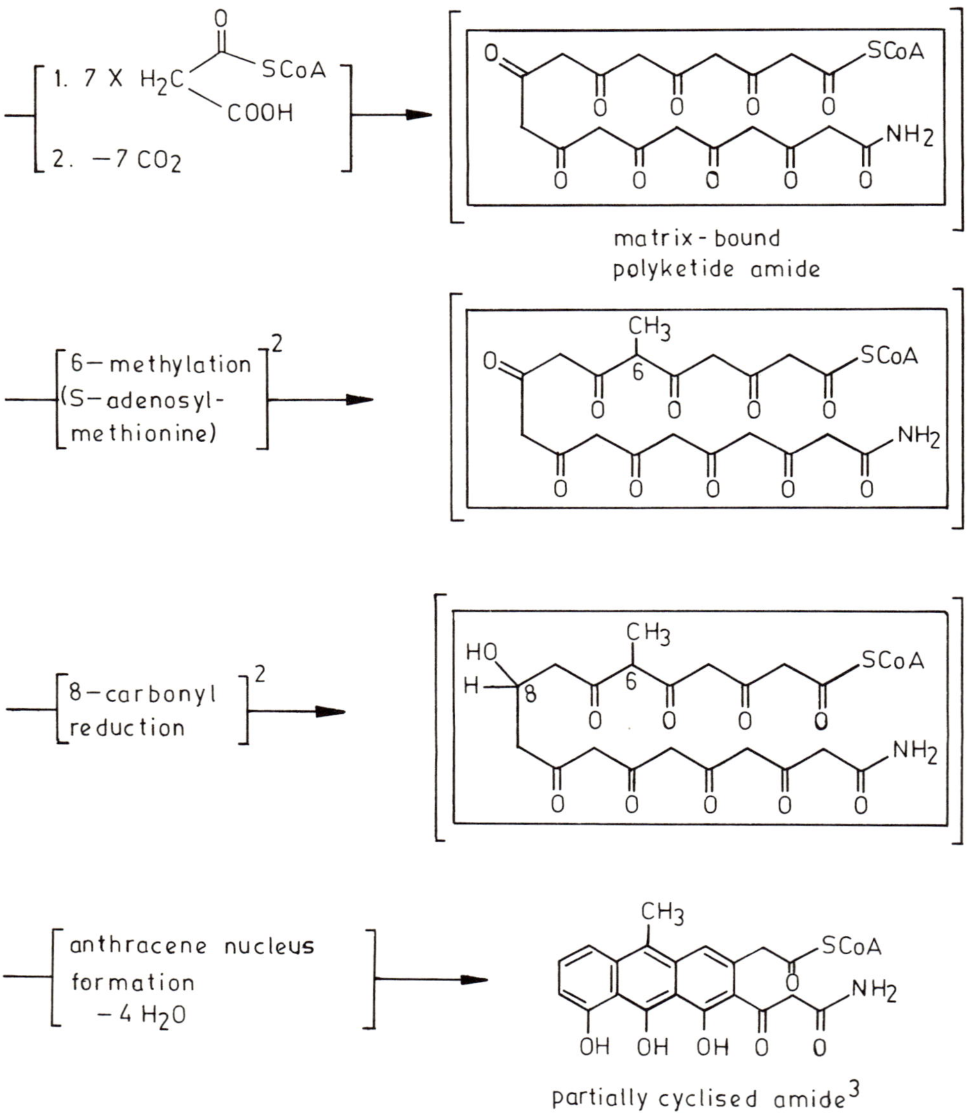
1. 7 X H_2C SCoA COOH
2. $-7\ CO_2$
SCoA
NH_2
matrix-bound
polyketide amide
6-methylation (S-adenosyl-methionine)[2]
CH_3
6
8-carbonyl reduction[2]
HO
H
8
anthracene nucleus formation
$-4\ H_2O$
OH OH OH
partially cyclised amide[3]

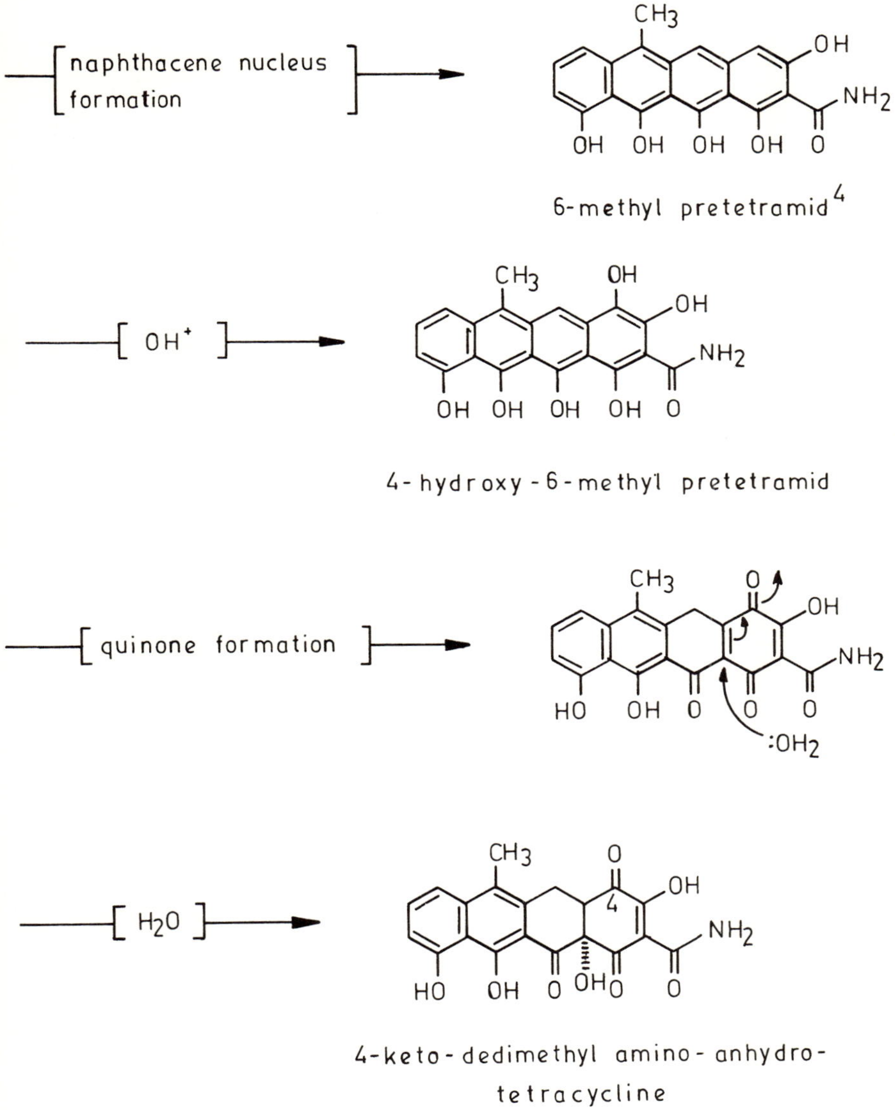
naphthacene nucleus formation
6-methyl pretetramid[4]
OH+
4-hydroxy-6-methyl pretetramid
quinone formation
:OH2
H2O
4-keto-dedimethyl amino-anhydro-tetracycline

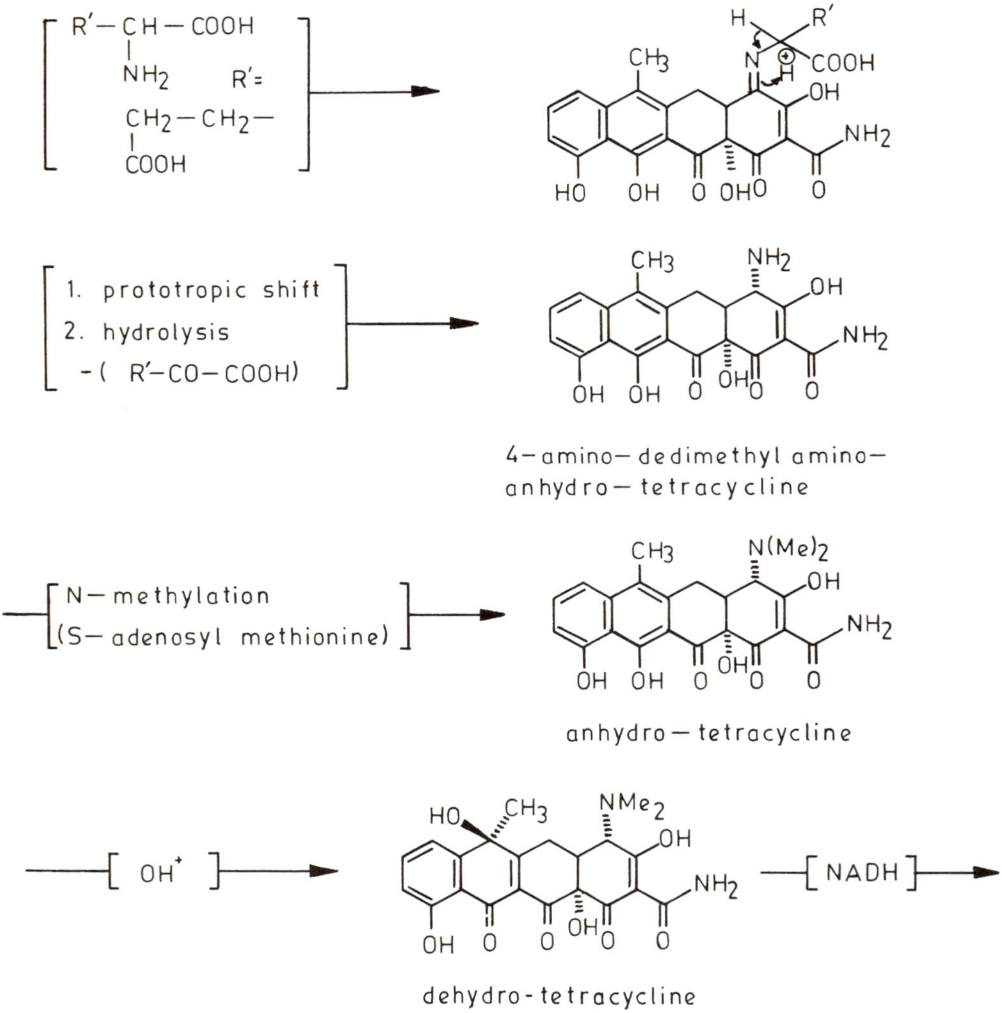
R′– CH – COOH
NH2
R′=
CH2 – CH2 –
COOH
1. prototropic shift
2. hydrolysis
-(R′– CO – COOH)
4-amino-dedimethyl amino-anhydro-tetracycline
N-methylation
(S-adenosyl methionine)
anhydro-tetracycline
OH+
dehydro-tetracycline
NADH

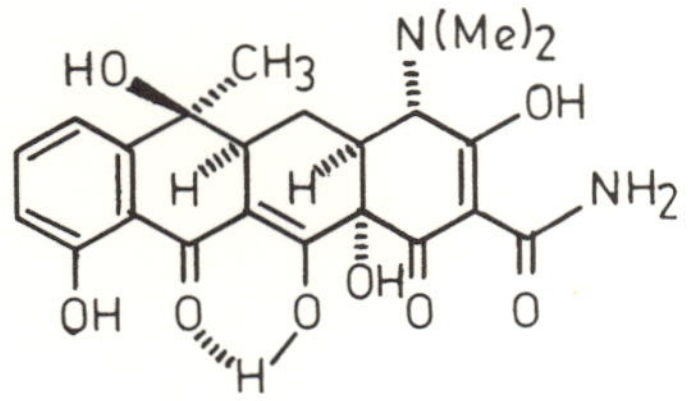

Tetracycline

---

1. J. R. D. McCormick in, 'Antibiotics', Vol. 2, eds. D. Gottlieb and P. D. Shaw, Springer-Verlag, New York, N.Y., 1967, p. 113; J. R. D. McCormick, E. R. Jensen, S. Johnson and N.O. Sjolander, J. Amer. Chem. Soc., 90, 2201 (1968); J. R. D. McCormick and E. R. Jensen, J. Amer. Chem. Soc., 91, 206 (1969).

2. This sequence of methylation and reduction was suggested on the basis of isolation and characterisation of products from blocked mutants at each of these steps (Ref. 1).

3. Methylanthrone, a shunt product of this intermediate has been isolated from a blocked mutant of S. aureofaciens.

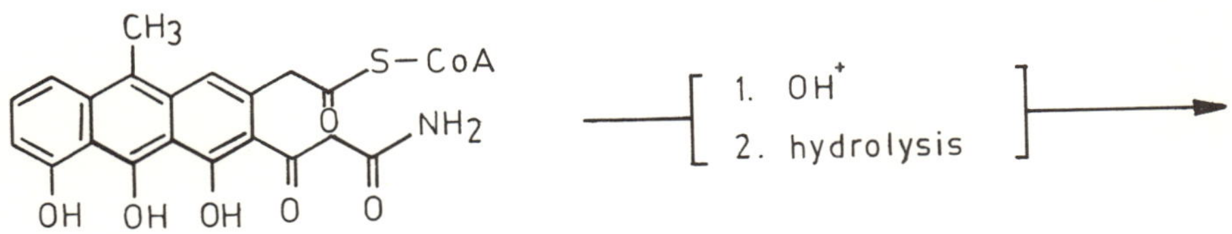

partially cyclised amide

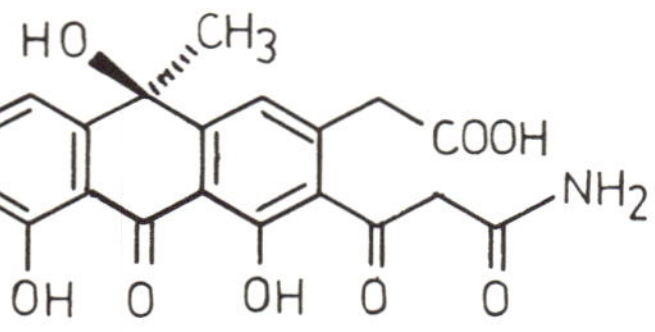

methyl anthrone

(J. R. D. McCormick, E. R. Jensen, N.H. Arnold, H. S. Corey, U. H. Joachim, S. Johnson, P. A. Miller and N. O. Sjolander, J. Amer. Chem. Soc., 90, 7127 (1968).)

4. 6-Methylpretetramid is biologically converted to tetracycline (J. R. D. McCormick, S. Johnson and N. O. Sjolander, J. Amer. Chem. Soc , 85, 1692 (1963)).

5. 4-Hydroxy-6-methyl-pretetramid accumulates in a blocked mutant of *S. aureofaciens* (J. R. D. McCormick, U. H. Joachim, E. R. Jensen, S. Johnson and N. O. Sjolander, J. Amer. Chem. Soc., 87, 1793 (1965)).

6. The observation of a dimethylamine-containing shunt product produced by the blocked mutant before reduction step shows that amination and dimethylation preceded anhydrotetracycline formation (Ref. 1).

7. Anhydrotetracycline is converted to tetracycline by a wide variety of blocked mutants of *S. aureofaciens* (Ref. 1).

TRYPTOPHAN

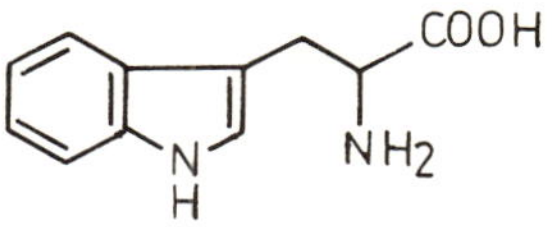

Tryptophan is Nature's favourite synthon in the creation of a very large number of natural products. The construction of Tryptophan itself is fascinating.[1]

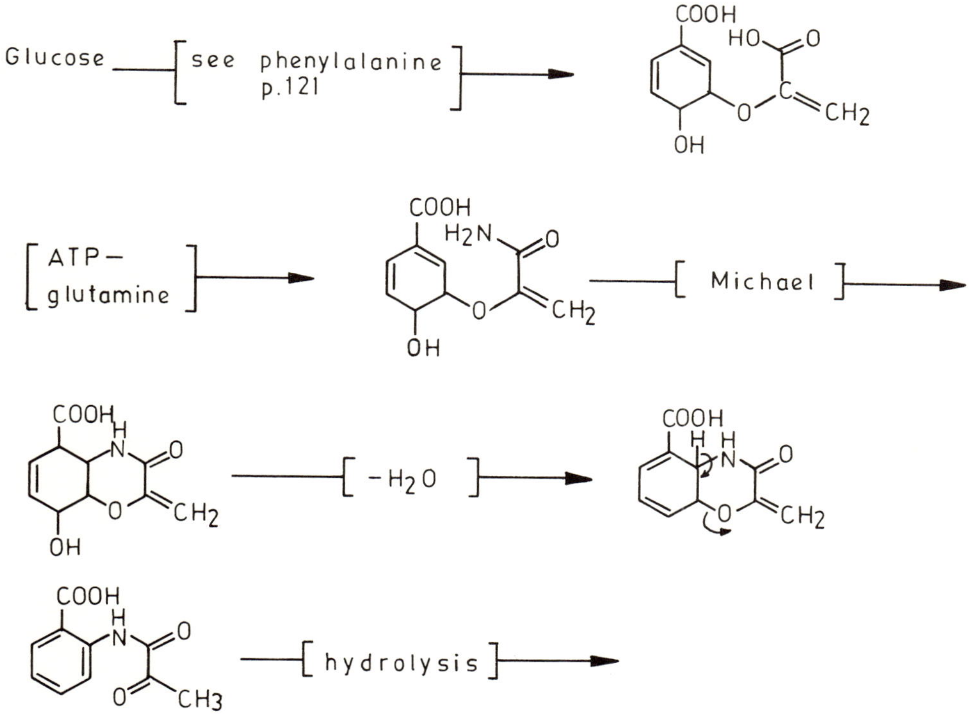

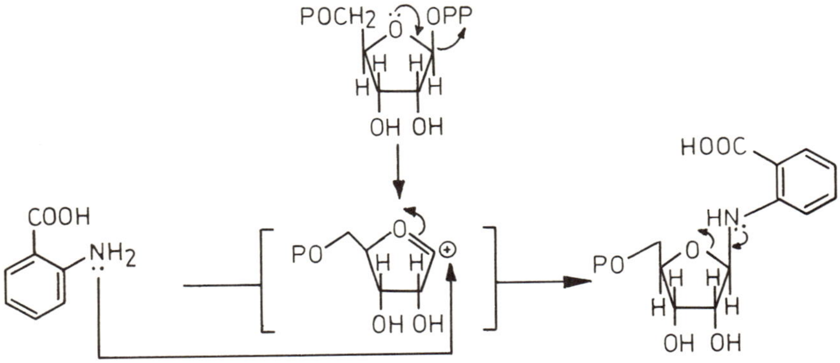

Anthranilic acid

N-(5'-phosphoribosyl)-anthranilic acid

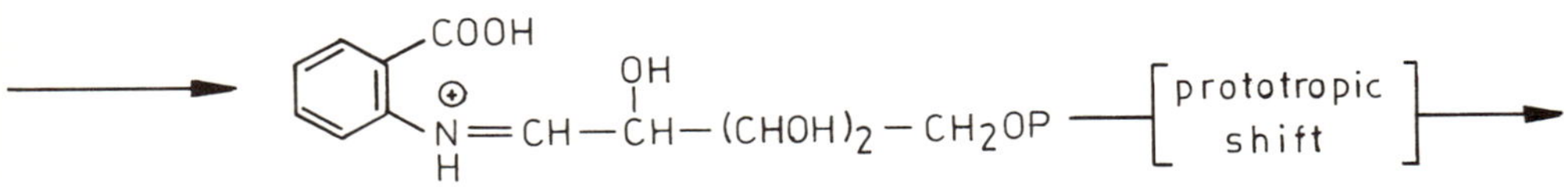

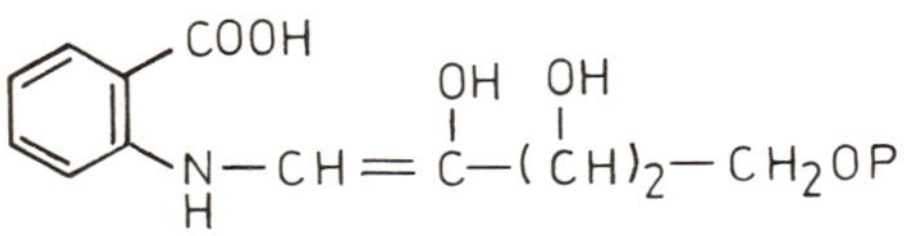

enol-1-(o-carboxyphenylamino)
1-deoxyribulose-5-phosphate

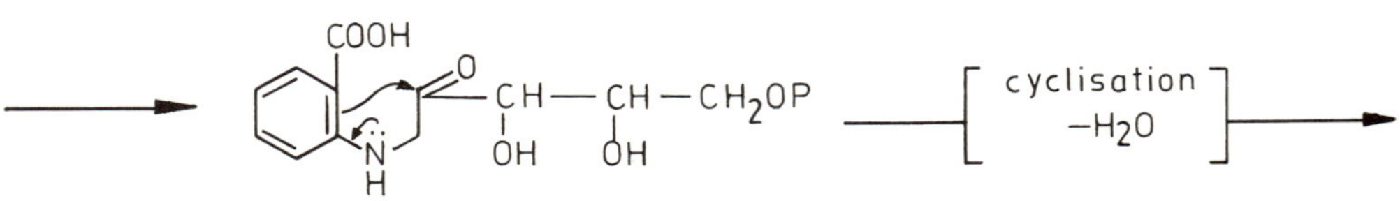

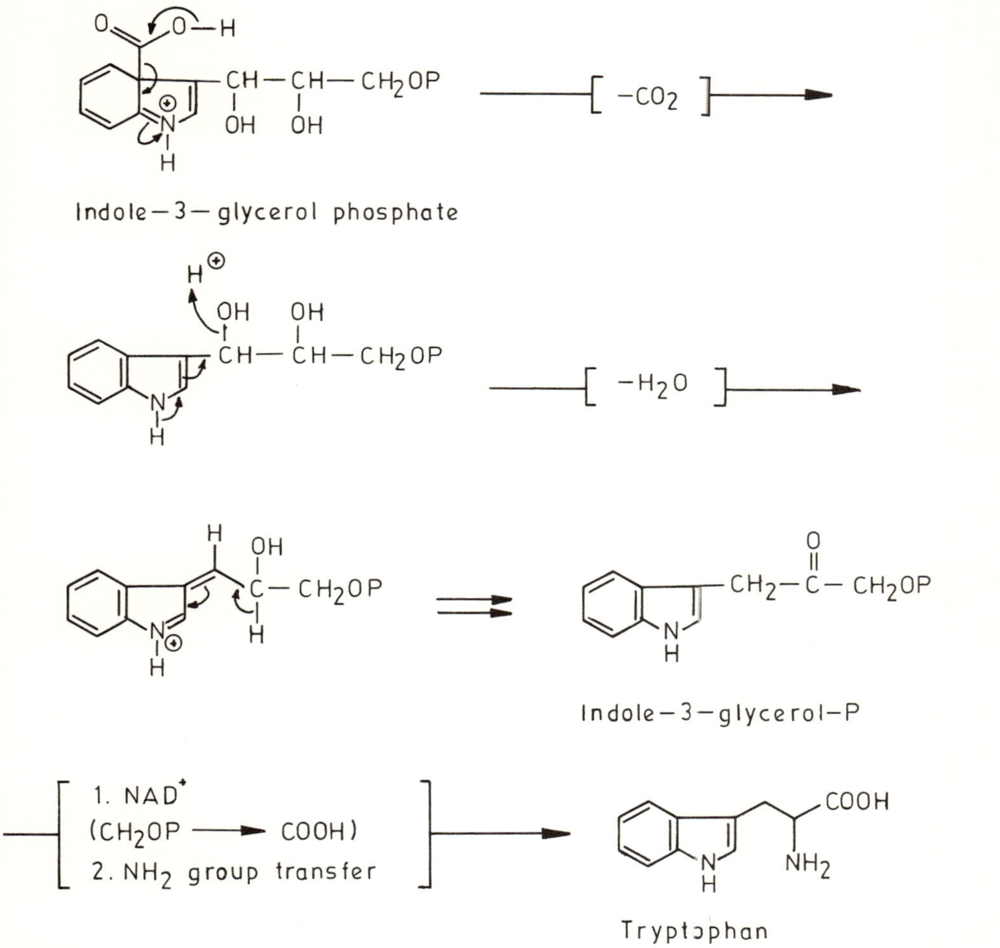

1. See, 'Biological Chemistry', Second Edition, Eds. H. R. Mahler and E. H. Cordes, Harper and Row, New York, 1971, p. 791.

USNIC ACID

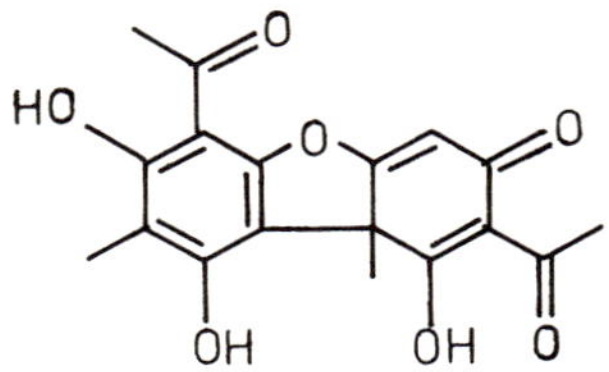

Barton's ingenious synthesis of usnic acid by phenolic coupling[1] was later shown to be precisely as practised by nature![2,3]

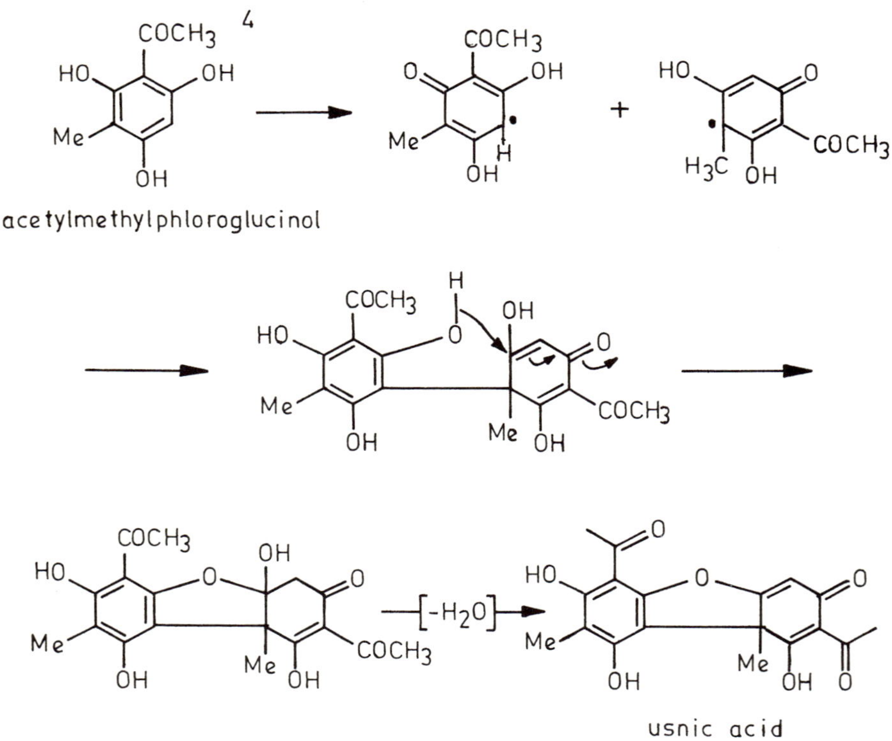

1. D. H. R. Barton, A. M. Deflorin and O. E. Edwards, J. Chem. Soc., 530 (1956).

2. A. Pentilla and H. M. Fales, Chem. Comm., 656 (1966).

3. H. Taguchi, U. Sankawa and S. Shibata, Tetrahedron Letters, 5211 (1966).

4. The acetylmethylphloroglucinol skeleton is derived biogenetically by the acetate-malonate condensation accompanying C_1-fragment incorporation.

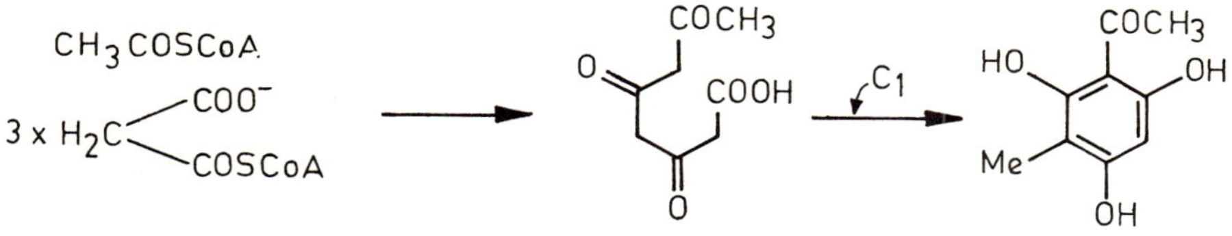

acetylmethlphloroglucinol

VINCOSIDE

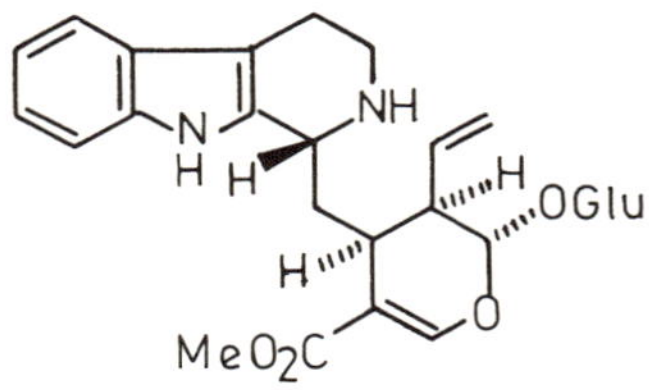

Vincoside, the fundamental unit in the biosynthesis of over 1000 alkaloids arises in Nature from the union of tryptamine and secologanin.[1,2]

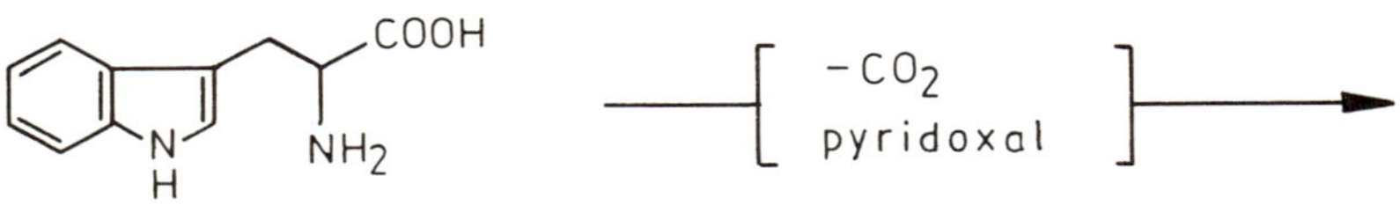

Tryptophan

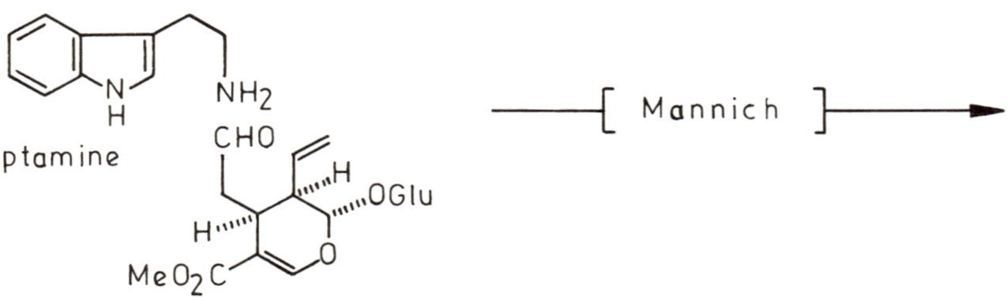

Tryptamine

Secologanin

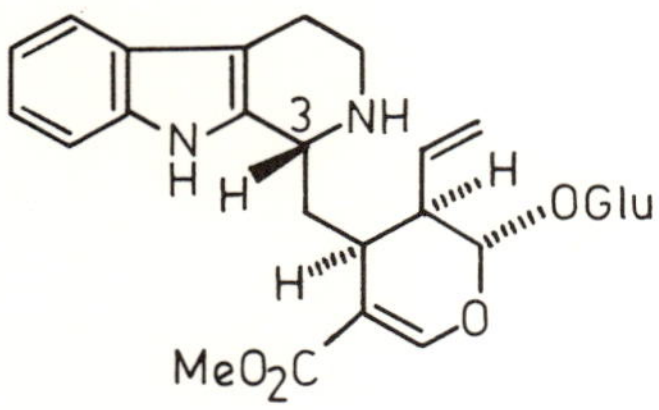

Vincoside[3]

---

1. A. I. Scott, Acc. Chem. Res., 3, 151 (1970).

2. The *in vitro* condensation of secologanin with tryptamine to form a separate mixture of vincoside and the 3-epimer, isovincoside has been achieved (A. R. Battersby, A. R. Burnett and P. G. Parsons, Chem. Comm., 1282 (1968); *ibid.*, J. Chem. Soc., (C), 1193 (1969); A. R. Battersby and R. J. Parry, Chem. Comm., 902 (1971)).

3. Vincoside undergoes inversion at C-3 at some stage during the biosynthesis of indole alkaloids which possess α -hydrogen at C-3. The C-3 label is retained in this process which can be rationalised as shown:

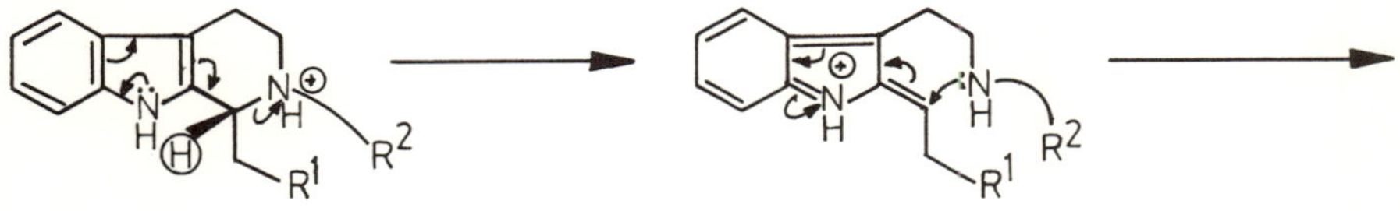

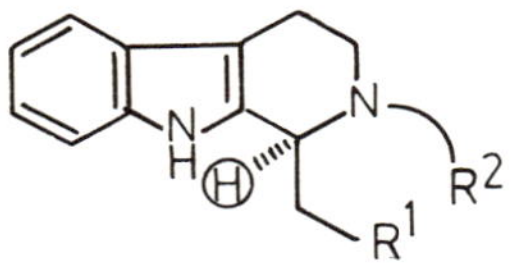

(A. I. Scott in 'Alkaloids' MTP International Review of Science, Organic Chemistry, Series one, Vol. 9, (eds.) D. H. Hey and K. F. Wiesner, Butterworths, University Park Press, London, 1973, p. 105).

VINDOLINE

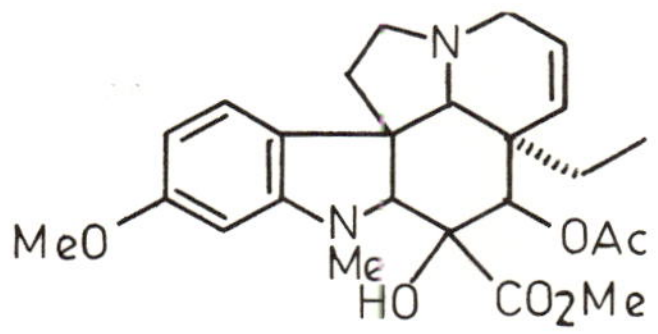

The biosyntheses of Catharanthine and Vindoline[1] from Vincoside are closely linked, the bifurcation point being the common intermediate I resulting from the fragnentation of Stemmadenine.

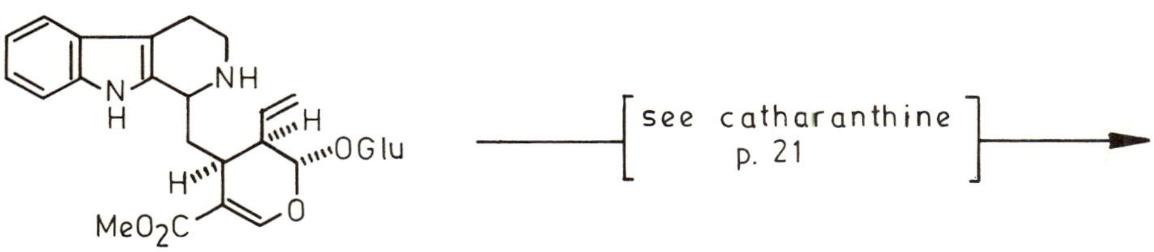

Vincoside

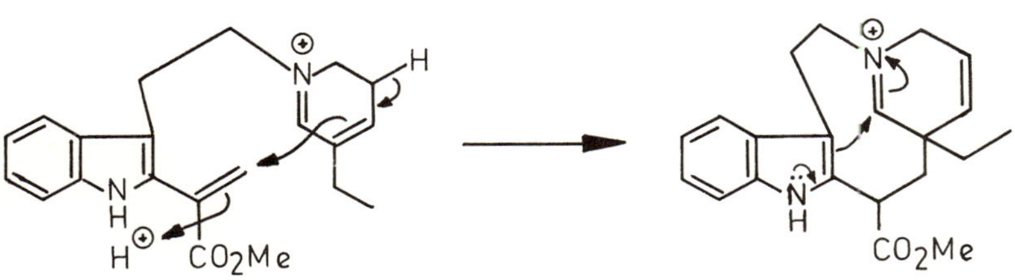

I

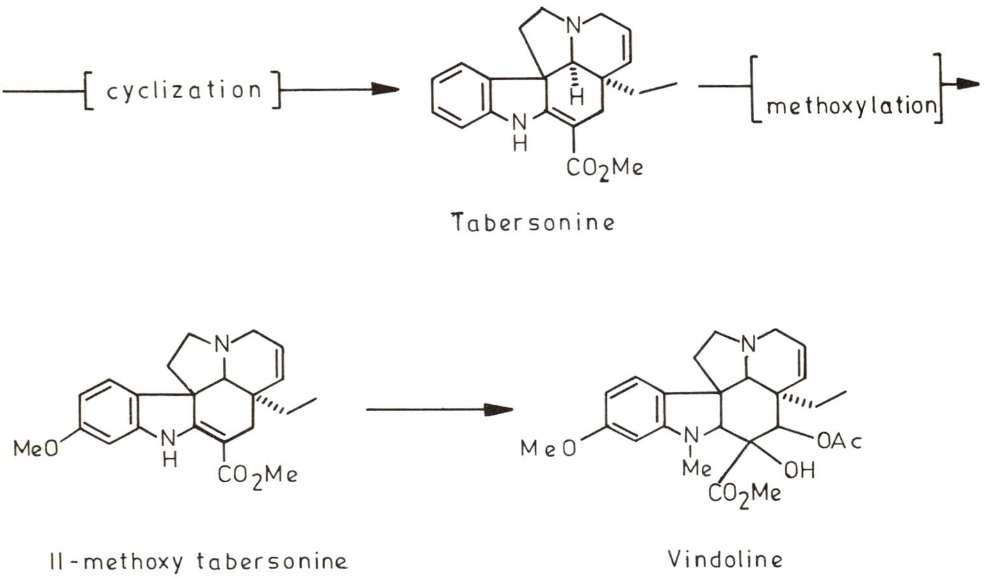

1. A. I. Scott, Acc. Chem. Res., 3, 151 (1970); A. R. Battersby, Acc. Chem. Res., 5, 148 (1972); A. I. Scott, Bioorg. Chem., 3, 398 (1974).

VITAMIN B_{12}

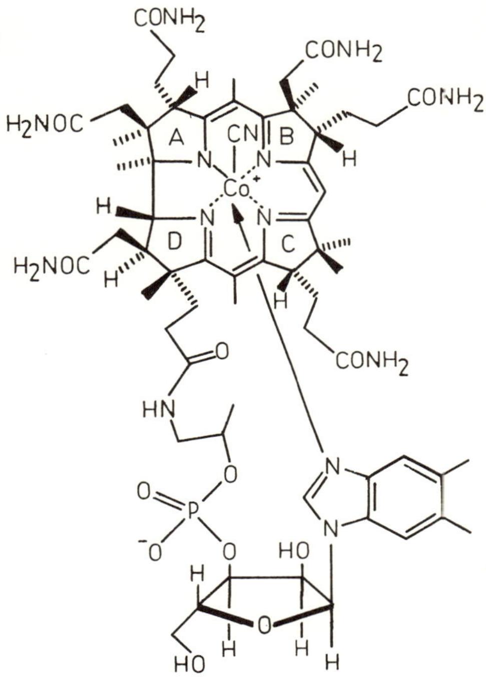

The synthesis of Vitamin-B_{12} was achieved after a decade long intensive and international assault.[1,2] Overlapping with this era - the sixties and early seventies - was a determined drive to understand - using new as well as classical techniques - how this molecule - which has presented so formidable an obstacle towards creation by man, arising mainly from a plethora of asymmetric centres that are present - is made in Nature.[3] The biosynthetic transformation of uroporphyrinogen III to Vitamin-

B_{12} makes one intensely aware of how advanced Nature is in the creation of molecular frameworks with a handful of coenzyme reagents.

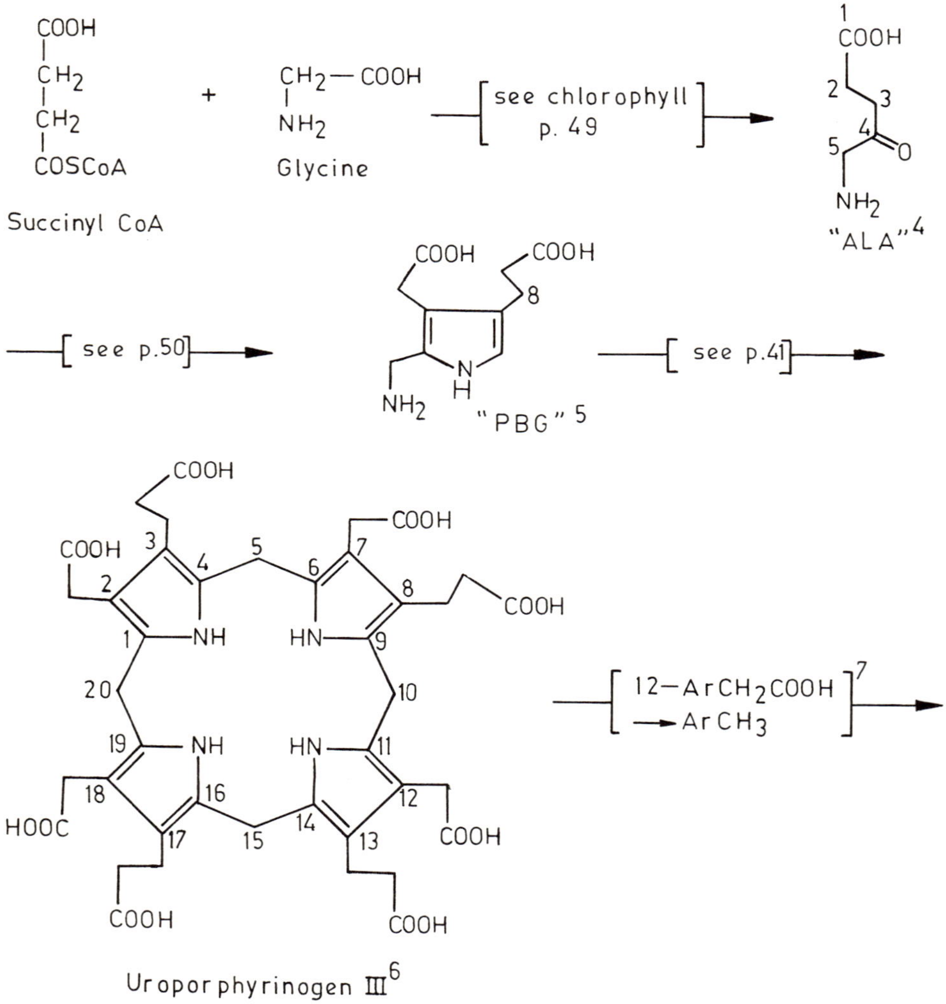

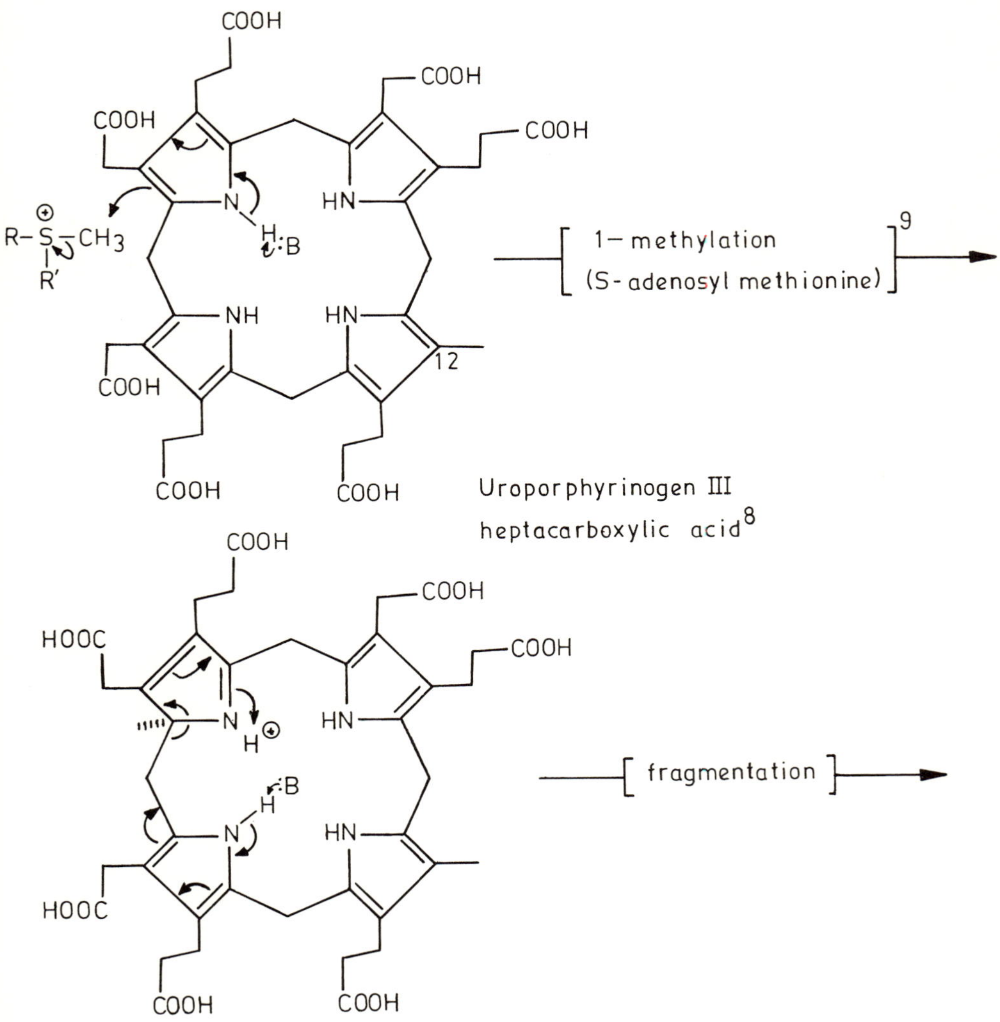

Uroporphyrinogen III heptacarboxylic acid[8]

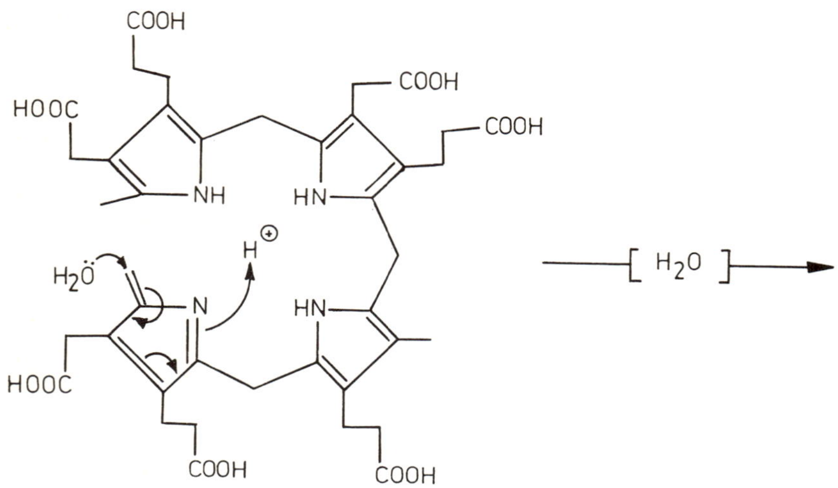
COOH
HOOC
COOH
COOH
NH
HN
H⊕
H2Ö
N
HN
HOOC
COOH
COOH
[H2O]

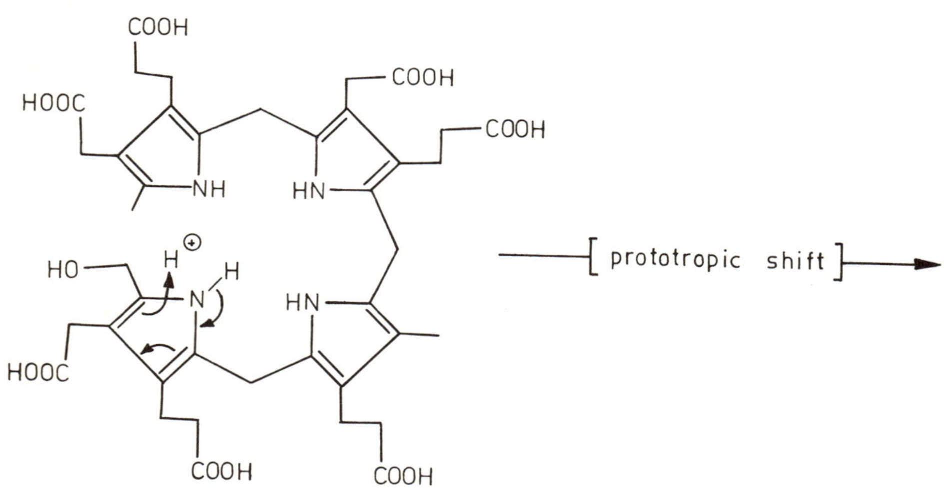
COOH
HOOC
COOH
COOH
NH
HN
H⊕
HO
N
H
HN
HOOC
COOH
COOH
[prototropic shift]

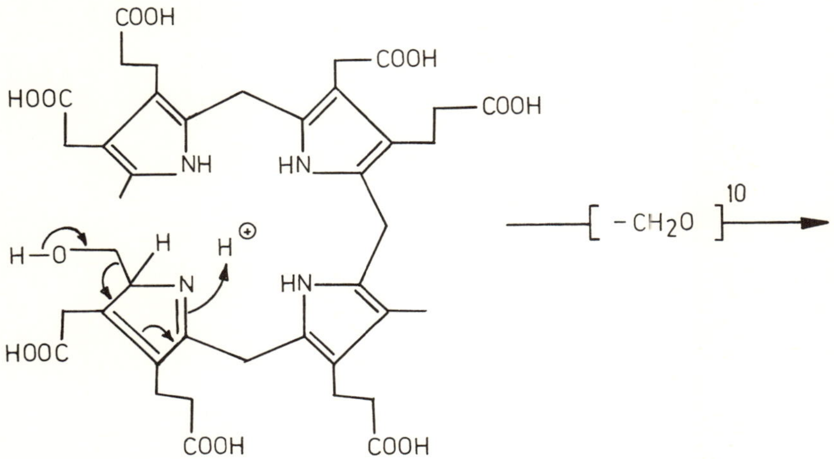
COOH
HOOC
NH
HN
H—O
H
N
HN
[-CH2O] 10

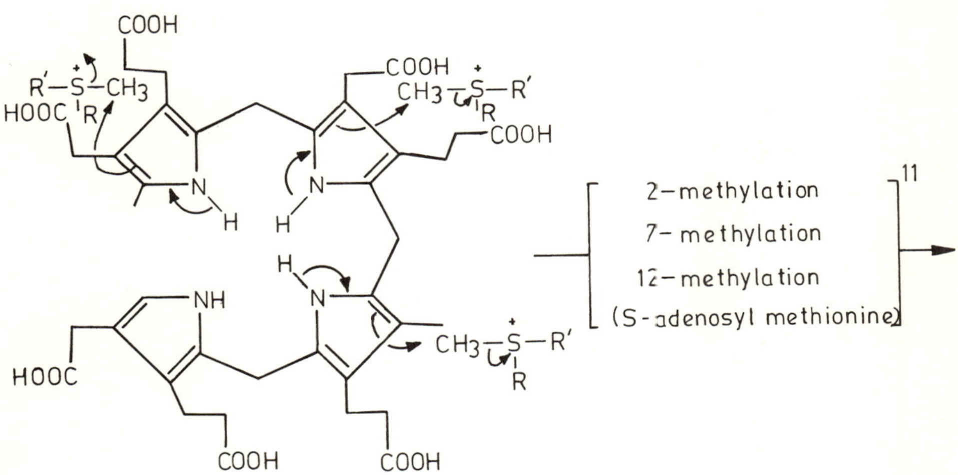
COOH
HOOC
R′—S—CH3
CH3—S—R′
R
N
H
NH
2–methylation
7– methylation
12–methylation
(S-adenosyl methionine)
11

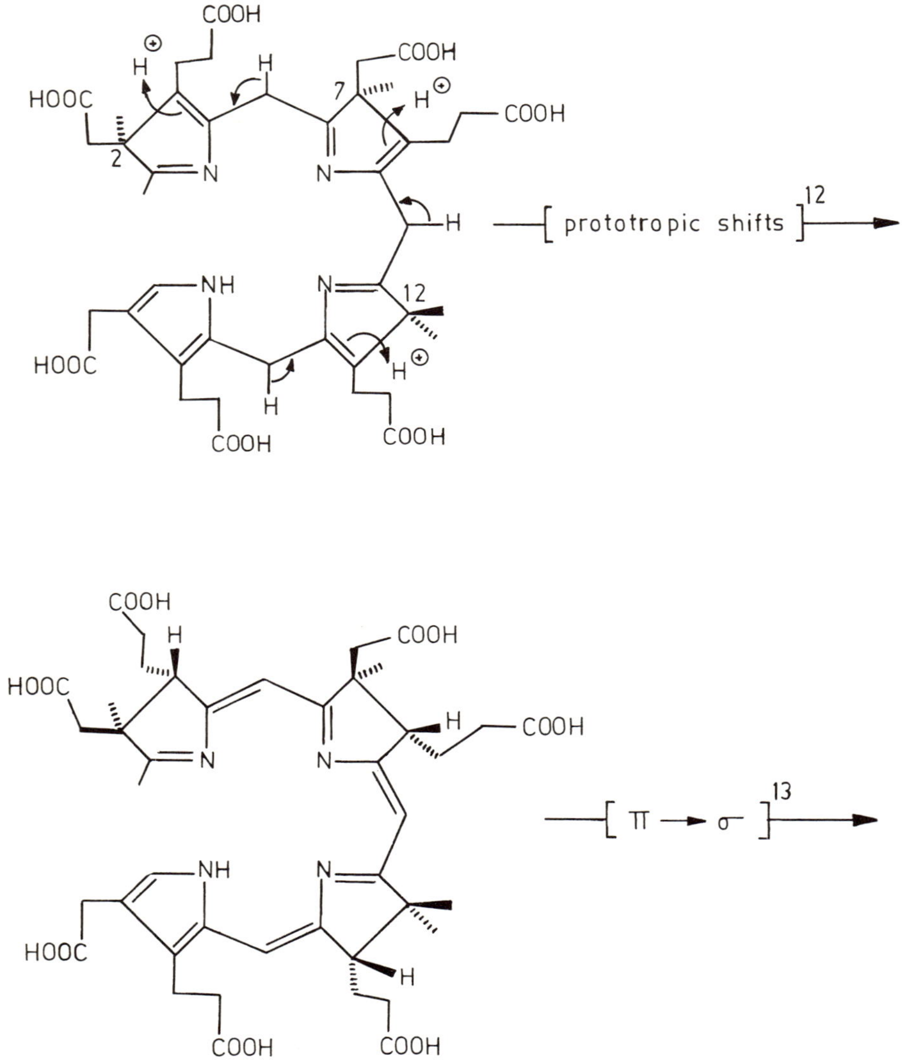
COOH
HOOC
H
2
7
12
NH
N
prototropic shifts
12
π → σ
13

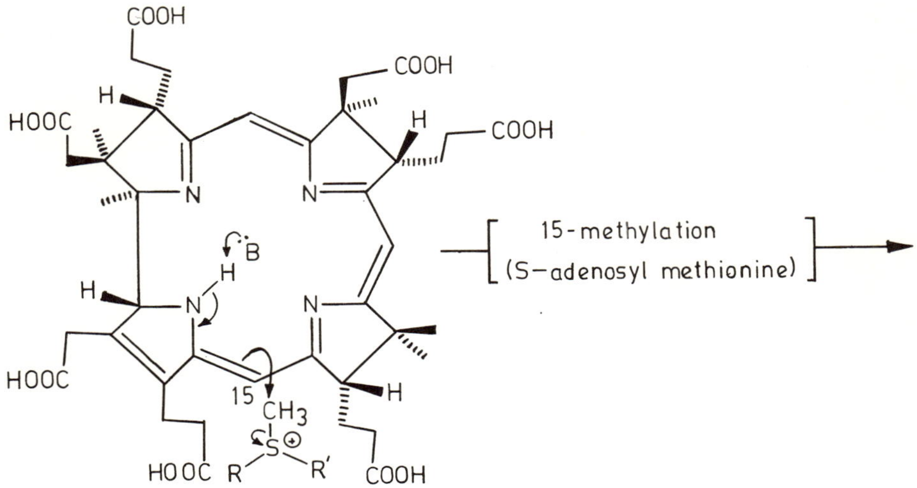
COOH
COOH
HOOC
H
H
COOH
N
N
15-methylation
(S-adenosyl methionine)
B
H
H
N
N
HOOC
15
CH3
H
S⊕
HOOC
R
R′
COOH

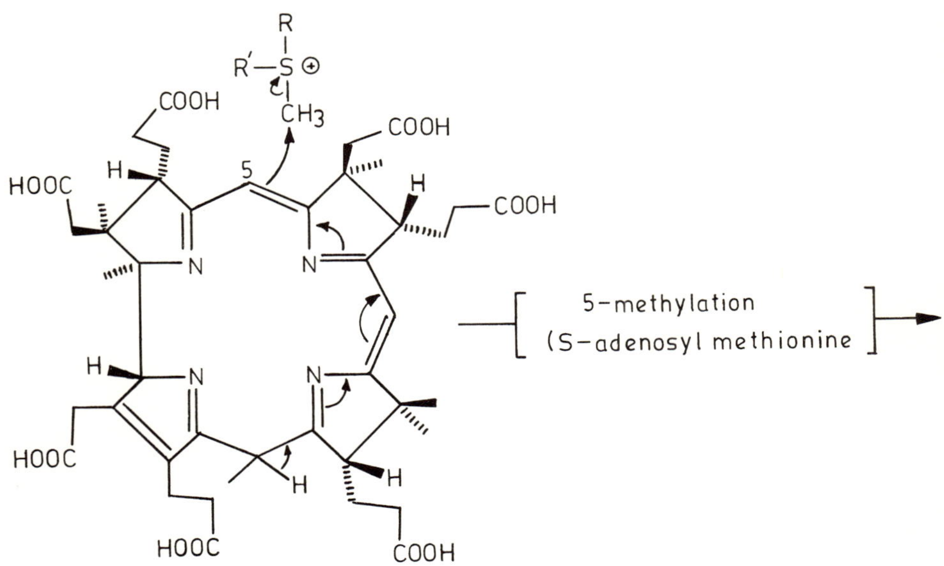
R
R′
S ⊕
CH3
COOH
COOH
H
5
H
HOOC
COOH
N
N
5-methylation
(S-adenosyl methionine
H
N
N
HOOC
H
H
HOOC
COOH

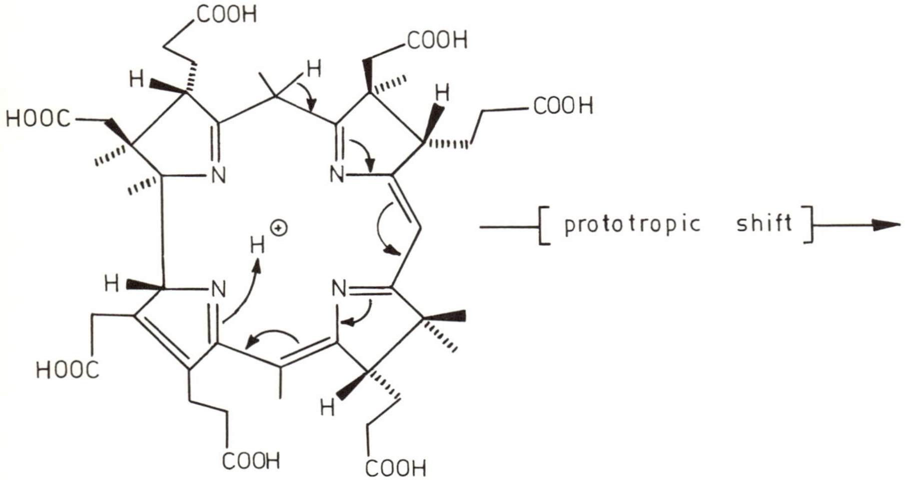
COOH
COOH
COOH
HOOC
H
H
H
N
N
H⊕
N
N
H
HOOC
H
COOH
COOH
prototropic shift

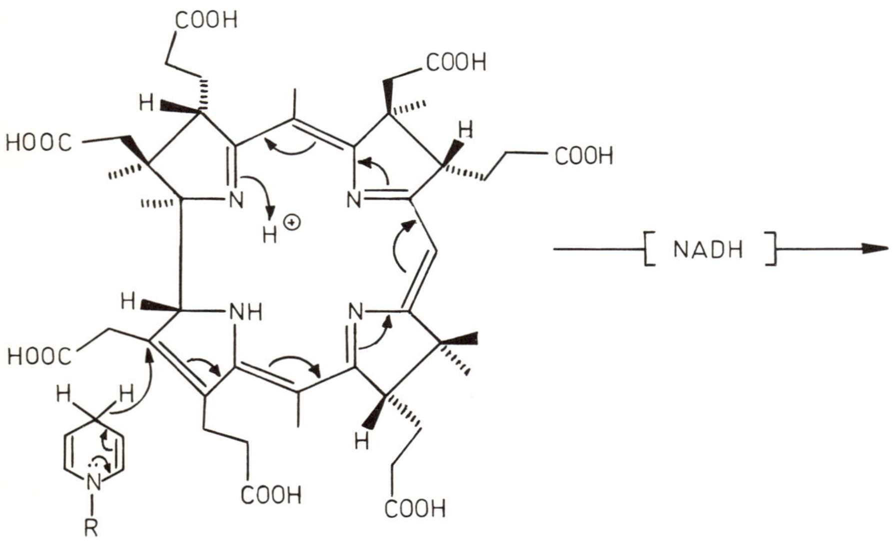
COOH
COOH
COOH
HOOC
H
H
N
N
H⊕
NH
N
H
HOOC
H
H
N
R
H
COOH
COOH
NADH

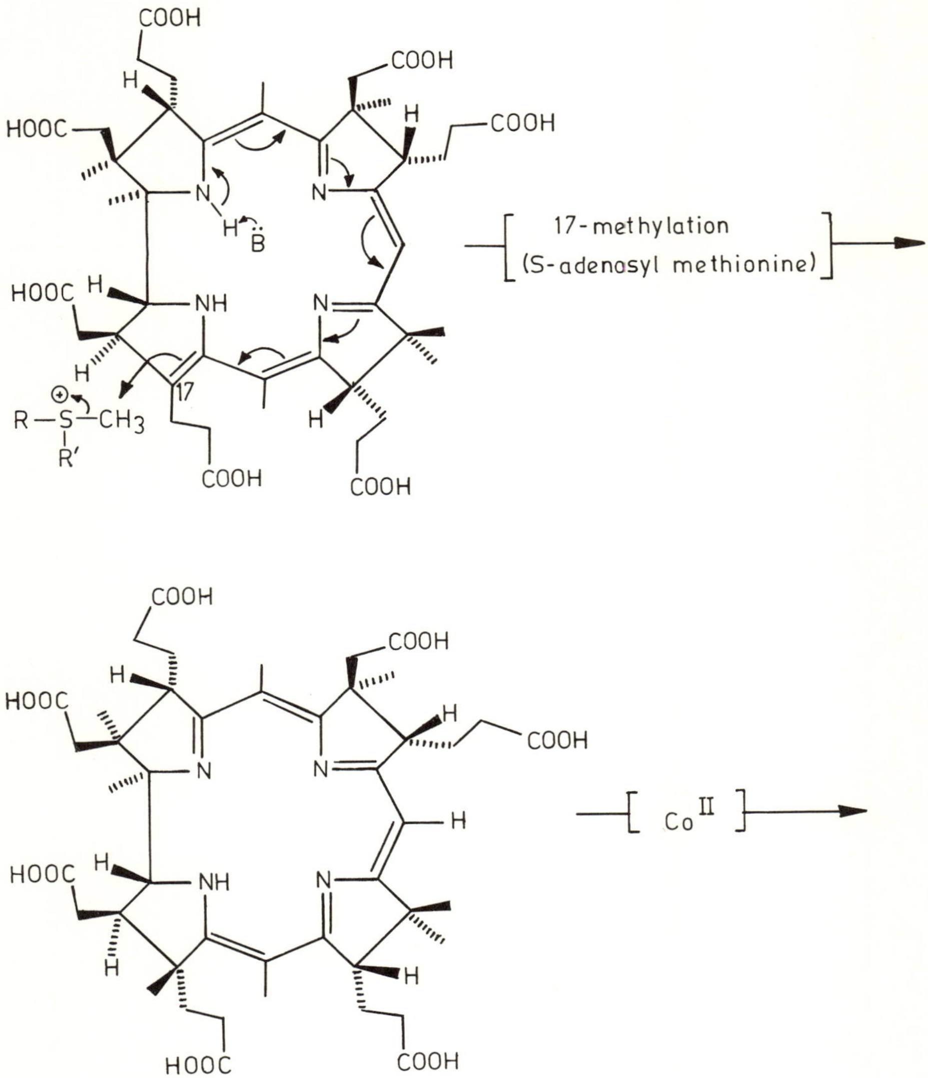
COOH
COOH
COOH
HOOC
H
N
N
H
B
17-methylation
(S-adenosyl methionine)
HOOC
NH
N
17
R—S—CH3
R′
COOH
COOH
COOH
COOH
COOH
HOOC
N
N
H
Co II
HOOC
NH
N
HOOC
COOH

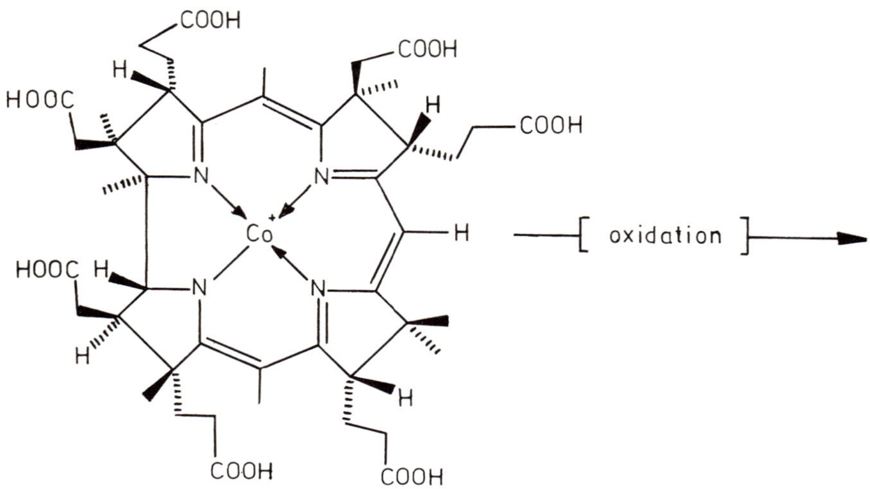

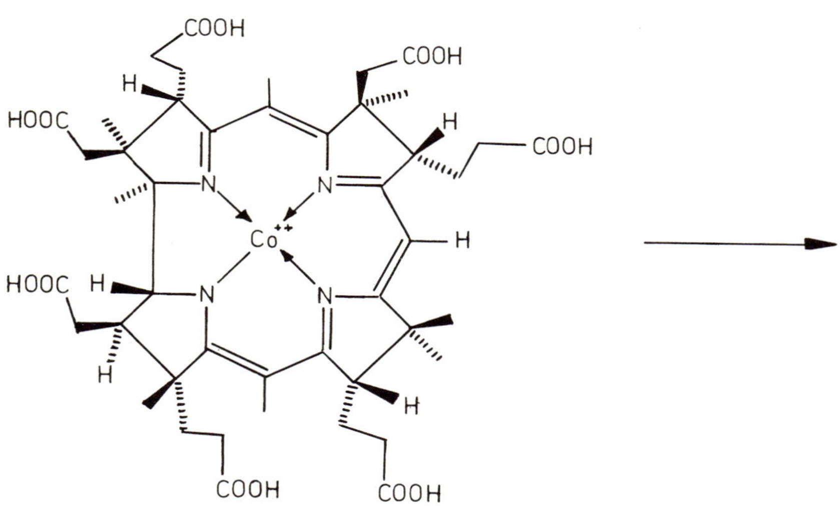

Cobyrinic acid[14]

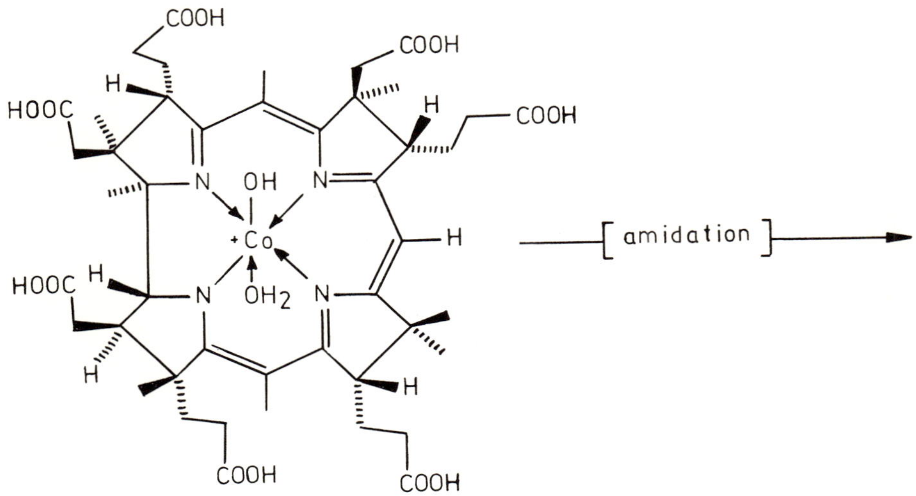

diaquo-cobyrinic acid

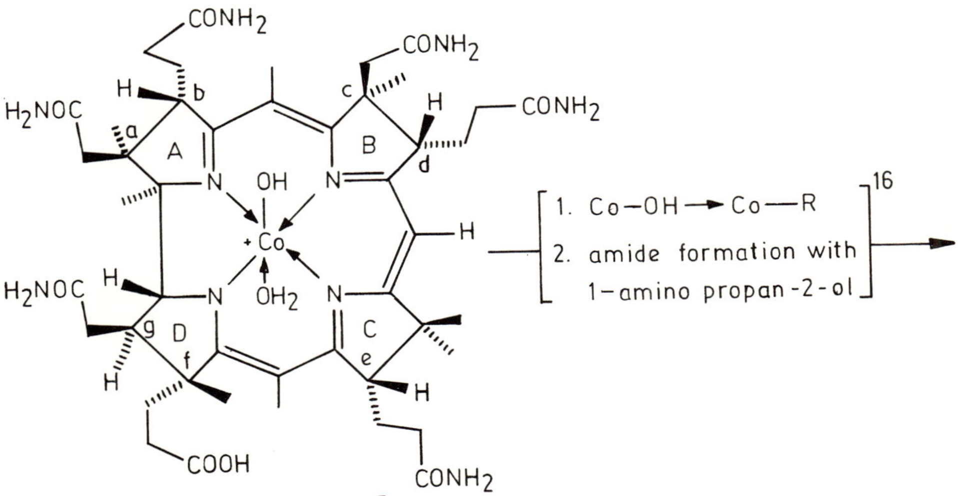

Cobyric acid[15]

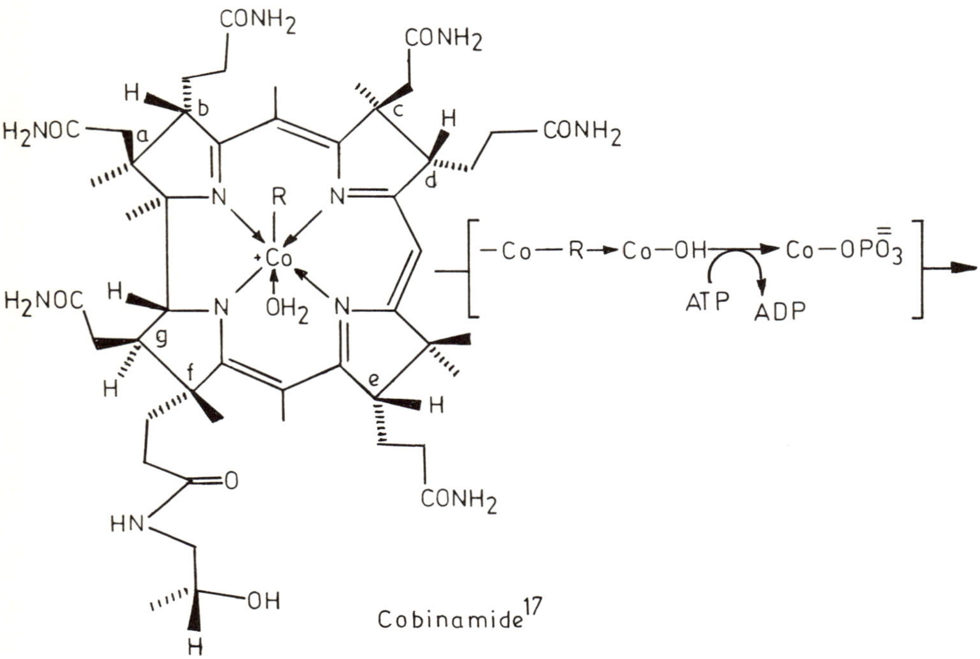

Cobinamide[17]

R =

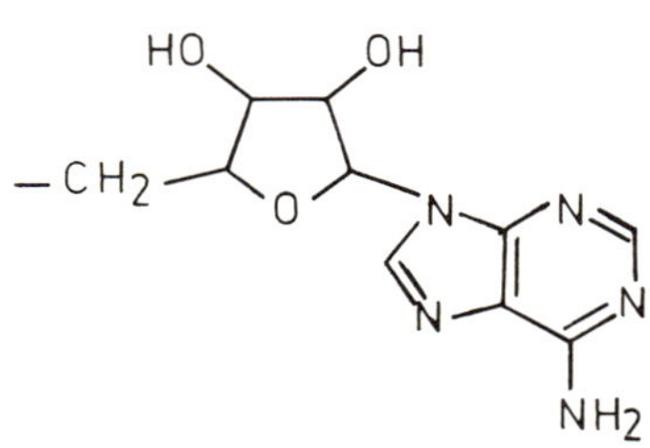

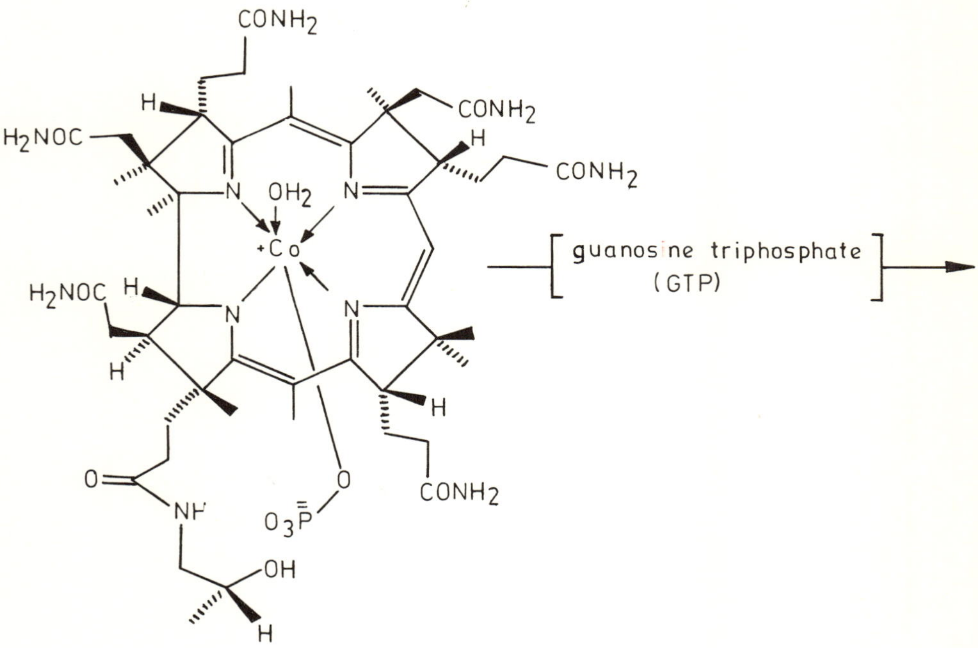

Cobinamide - phosphate[18]

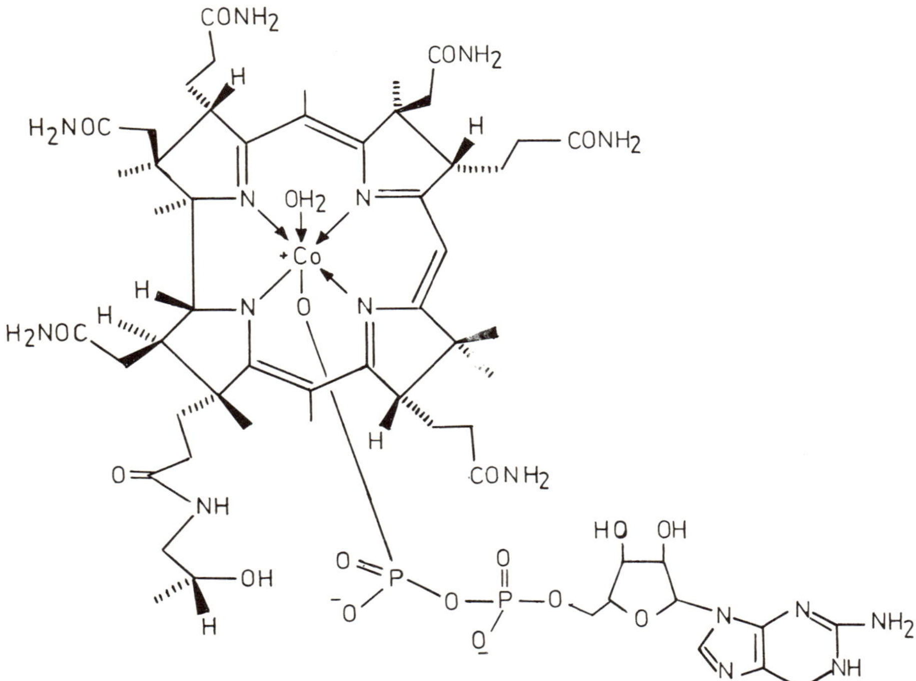

Cobinamide Guanosine diphosphate[19]

[20,21,22]

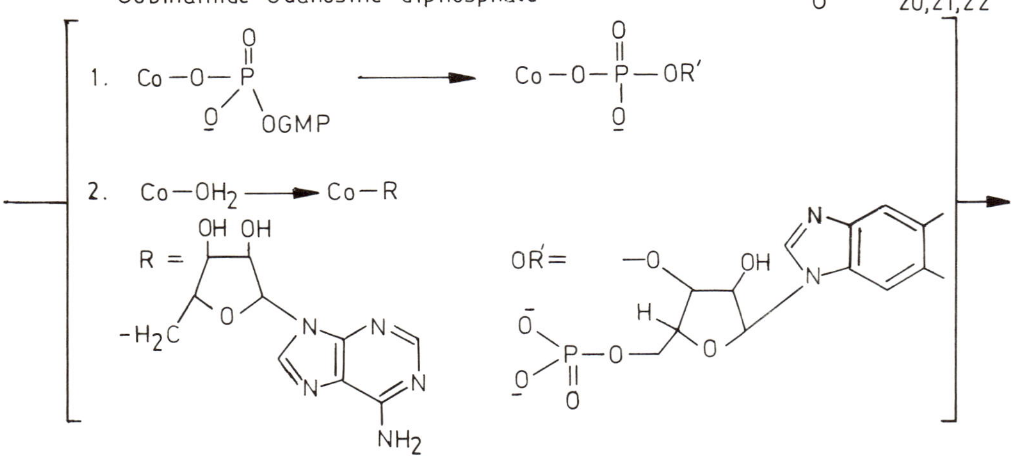

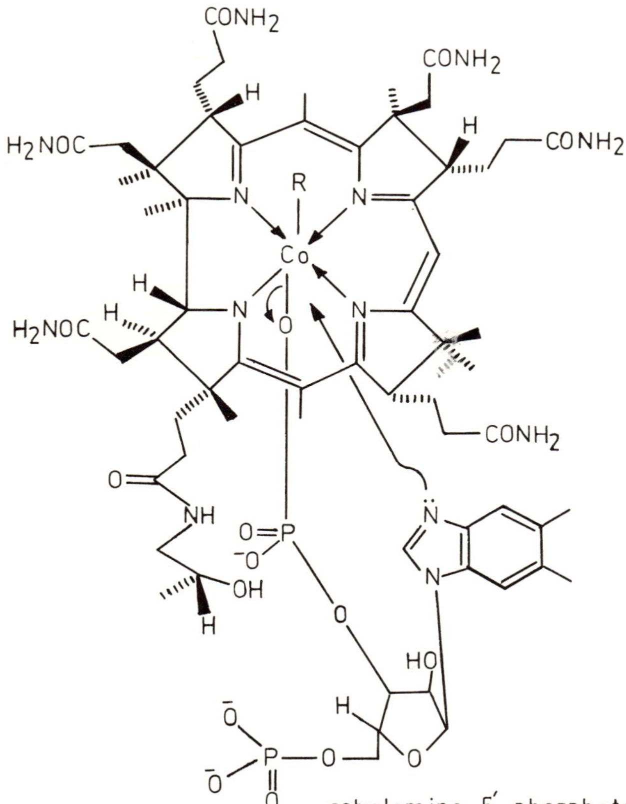

cobalamine-5′-phosphate

1. imidazole Co-bond formation
2. hydrolysis of ribose-5′-phosphate
3. phosphate ester formation linking ribazole and D-ring side chain

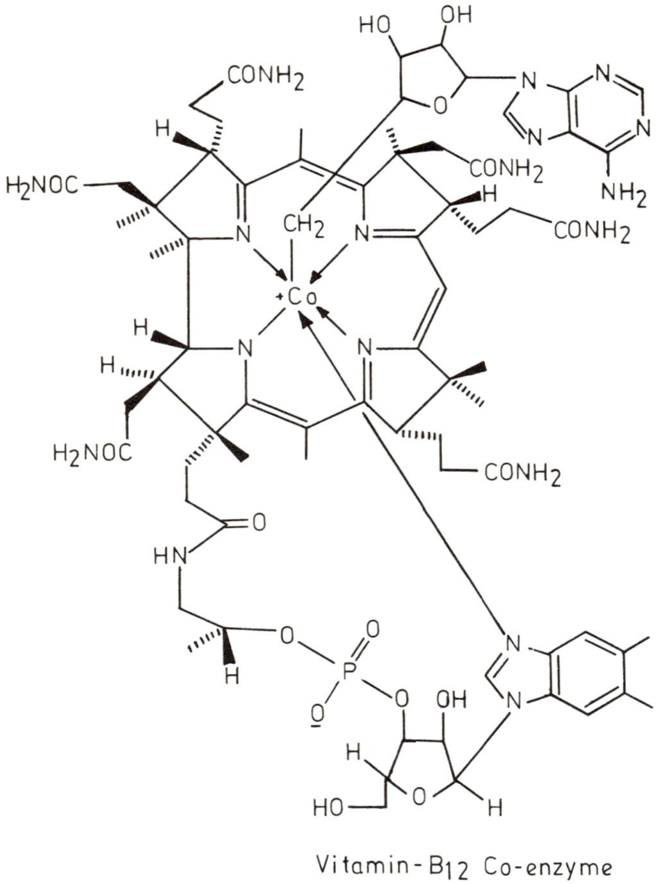

Vitamin-B_{12} Co-enzyme

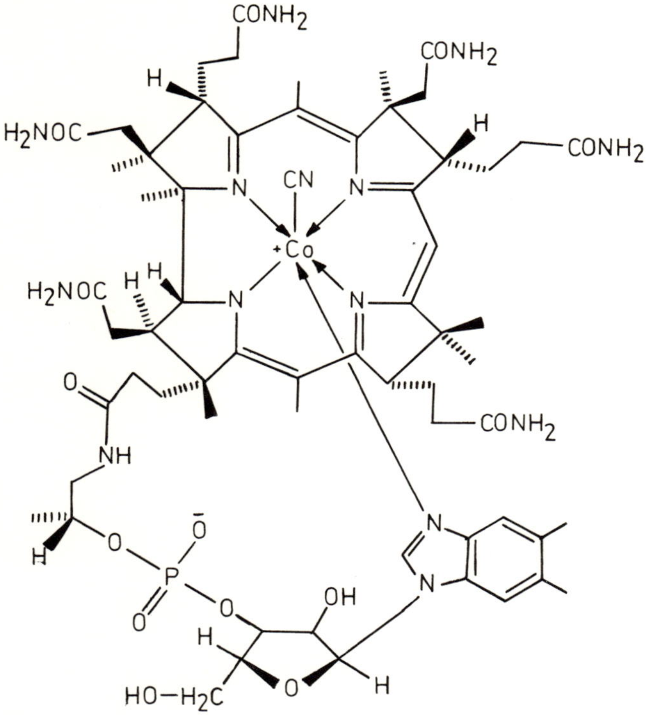

Vitamin - B_{12}

1. R. B. Woodward, IUPAC 'Chemistry of Natural Products', Vol.VIII, (New Delhi), Butterworths, London, 1972, p. 145; Pure Appl. Chem., 17, 519 (1968); A. Eschenmoser, IUPAC '23rd Congress, Special Lectures', Vol. II (Boston), Butterworths, London, 1971, p. 69; Quart. Rev., Chem. Soc., 24, 366 (1970).

2. For a comprehensively referenced article on vitamin-B_{12} see, P. Dowd, M. Shapiro and K. Kang, J. Amer. Chem. Soc., 97, 4754 (1975).

3. B. F. Burnham in 'Metabilic Pathways', Vol. III, 3rd ed., D. M. Greenberg, Ed., Academic Press, New York, N.Y., 1969, Chapter 18; D. Shemin, J. W. Corcoran, C. Rosenblum and I. M. Miller, Science, 124, 272 (1956); J. W. Corcoran and D. Shemin, Biochim. Biophys. Acta, 25, 661 (1957); R. C. Bray and D. Shemin, J. Biol. Chem., 238, 1501 (1963); R. C. Bray and D. Shemin, Biochim. Biophys. Acta, 30, 647 (1958); A. J. Krasna, C. Rosenblum and D. B. Sprinson, J. Biol. Chem., 225, 745 (1957); W. L. Alworth, S. H. Lu and M. F. Winkler, Biochemistry, 10, 1421 (1971); K. Bernhauer, F. Wagner, M. Heidrun, P. Rapp and H. Vogelmann, Hoppe-Seyler's, Z. Physiol. Chem., 349, 1297 (1968); K. Bernhauer, H. Vogelmann and F. Wagner, ibid ., 349, 1271, 1281 (1968), and previous papers in this series; F. Wagner, Annu. Rev. Biochem., 35, 405 (1966); H. C. Friedmann and L. M. Cagen, Annu. Rev. Microbiol., 24, 159 (1970); A. I. Scott, C. A. Townsend, K. Okada and M. Kajiwara, J. Amer. Chem. Soc., 96, 8054 (1974); A. I. Scott, C. A. Townsend, K. Okada, M. Kajiwara, R. J. Cushley and P. J. Whitman, J. Amer. Chem. Soc., 96, 8069 (1974).

4. Excellent specific incorporation of [2-^{13}C] ALA, [5-^{13}C] ALA into cyanocobalamin has been demonstrated using resting cells of *Propionibacterium sheramanii* (A. I. Scott. et al. in Ref. 3).

5. [8-^{13}C] PBG has been incorporated into cyanocobalamin using whole cell preparations of *P. shermanii* (A. I. Scott, C. A. Townsend, K. Okada, M. Kajiwara and P. J. Whitman, J. Amer. Chem. Soc., 94, 8267 (1972); A. I. Scott, C. A. Townsend, K. Okada and M. Kajiwara, J. Amer. Chem. Soc., 94, 8269 (1972); and A. I. Scott et al. in Ref. 3).

6. Recent intact incorporations of labelled [α,γ-$^{13}C_2$] and double labelled [α,γ-$^{14}C_2$] and [$^3H_2$](in ring B propionic acid) uroporphyrinogen III into cyanocobalamin and cobester using whole cells of *P. shermanii* (A. I. Scott, N. Georgopapadakou, K. S. Ho, S. Klioze, E. Lee, S. L. Lee, G. H. Temme III, C. A. Townsend and I. M. Armitage, J. Amer. Chem. Soc., 97, 2548 (1975); A. R. Battersby, M. Ihara, E. McDonald, F. Satoh and D. C. Williams, Chem. Comm., 436 (1975)) provided unequivocal evidence for the intermediacy of uroporphyrinogen III in the biosynthesis of vitamin-B_{12}.

7. The earlier suggestion (R. C. Bray and D. Shemin, J. Biol. Chem., 238, 1501 (1963)) that the β-methyl at C-12 arises

by the decarboxylation of an acetic acid residue, *i.e.* the C-2 of ALA, has now been unequivocally proved by ^{13}C-FT nmr analysis:

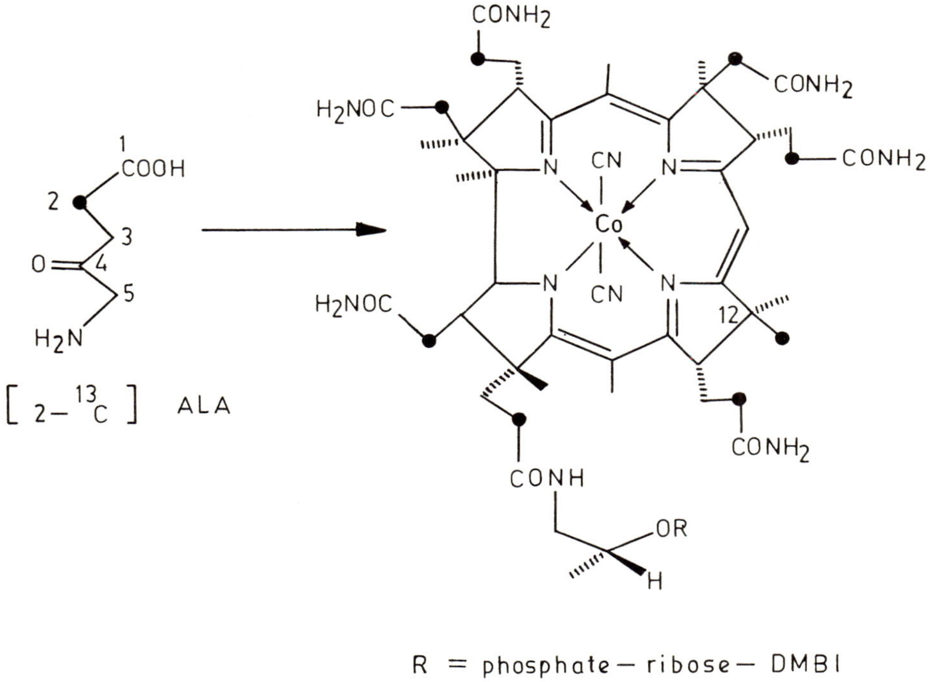

This loss of carbon dioxide must take place before C-12 methylation (*vide infra*), since the driving force for decarboxylation would be then lost:

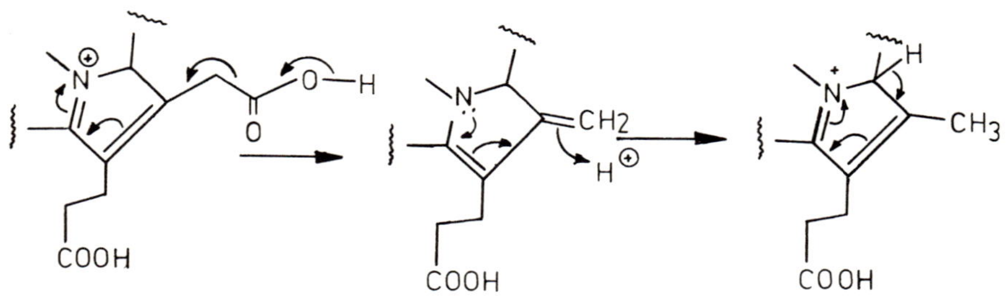

(A. I. Scott, C. A. Townsend, K. Okada, M. Kajiwara, P. J. Whitman and R. J. Cushley, J. Amer. Chem. Soc., 94, 8267 (1972)).

8. Recent excellent incorporation of [α,γ-$^{14}C_2$] heptacarboxylic uroporphyrinogen III into cobester supports the initial decarboxylation at C-12. (A. I. Scott, N. Georgopapadakou, K. S. Ho, S. Klioze, E. Lee, S. L. Lee, G. H. Temme III, C. A. Townsend and I. M. Armitage, J. Amer. Chem. Soc., 97, 2548 (1975).)

9. Feeding experiments with [$^{13}CH_3$]-methinine showed that out of the eight methyl groups on the peripheri of corrin nucleus, seven methyls ($C_1(\alpha)$, $C_2(\alpha)$, C_5, $C_7(\alpha)$, $C_{12}(\alpha)$, C_{15} and $C_{17}(\beta)$) are derived from methionine (A. I. Scott, C. A. Townsend, K. Okada, M. Kajiwara, P. J. Whitman and R. J. Cushley, J. Amer. Chem. Soc., 94, 8267 (1972); A. I. Scott, C. A. Townsend and R. J. Cushley, J. Amer. Chem. Soc., 95, 5759 (1973); A. R. Battersby, M. Ihara, E. McDonald, J. R. Stephenson and B. T. Golding, Chem. Comm., 404 (1973)).

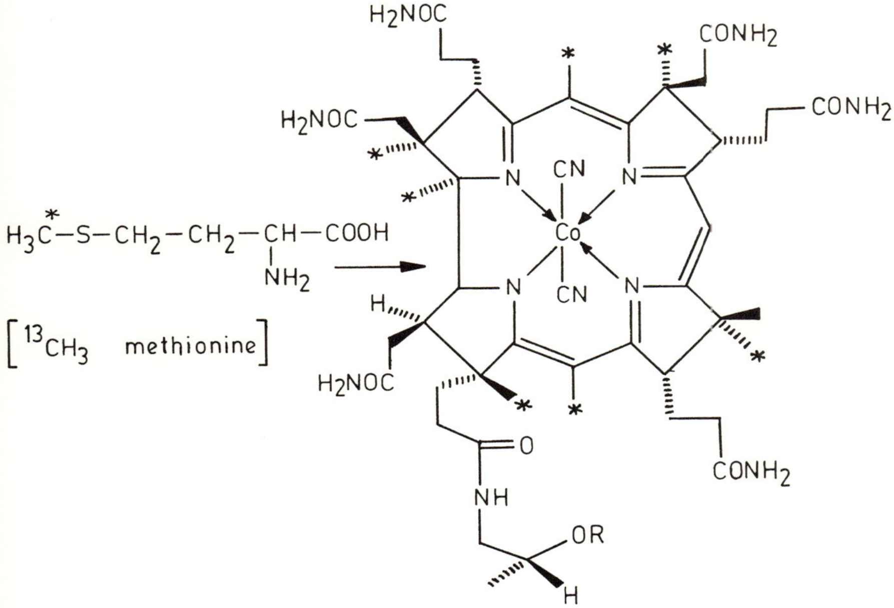

R = phospate – ribose – DMBI

10. One of the most obvious differences between the porphyrin framework - present in chlorophyll, Haemoglobin, Cytochromes - and the corrin skeleton of Vitamin-B_{12} is the absence of the methylene bridge linking A-D rings. Indeed, many ingenious explanations were advanced for the loss of this bridge, amongst which the most popular was its transformation to corrin-1-methyl as exemplified by the following rationalisation:

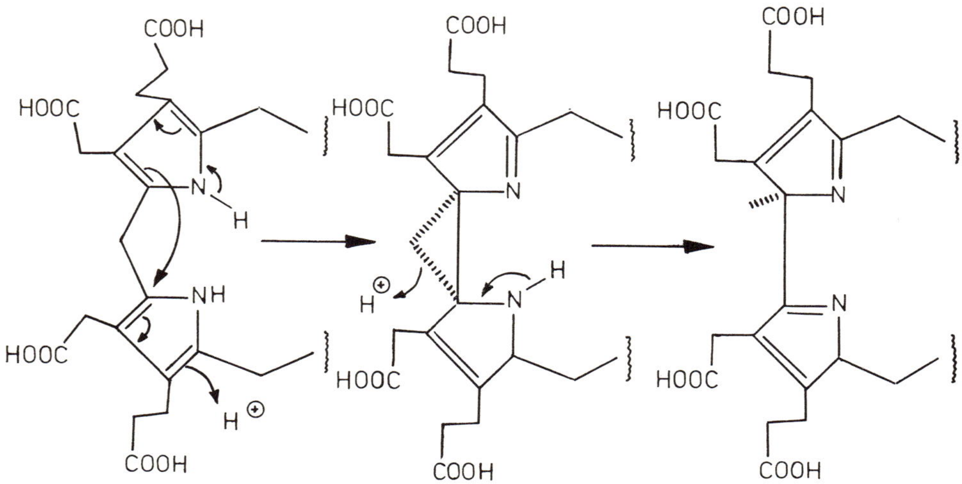

(A. I. Scott, C. A. Townsend, K. Okada and M. Kajiwara, Trans. N.Y. Acad. Sci., 35, 72 (1973)).

Such and similar rationalisations (J. H. Mathewson and A. H. Corwin, J. Amer. Chem. Soc., 83, 135 (1961)) would warrant that the C-1 methyl group originates from C_5 of ALA, which turns out not to be the case since this methyl has been clearly shown to originate from S-adenosyl methionine (see Ref. 9). Additionally, feeding experiments with [5-^{13}C] ALA showed that out of the eight anticipated enriched carbons, only seven appeared in the low field region implying that one of the -$^{13}CH_2$-NH_2 termini of ALA (and hence of PBG or uroporphyrinogen III) has been lost in the formation of corrin nucleus.

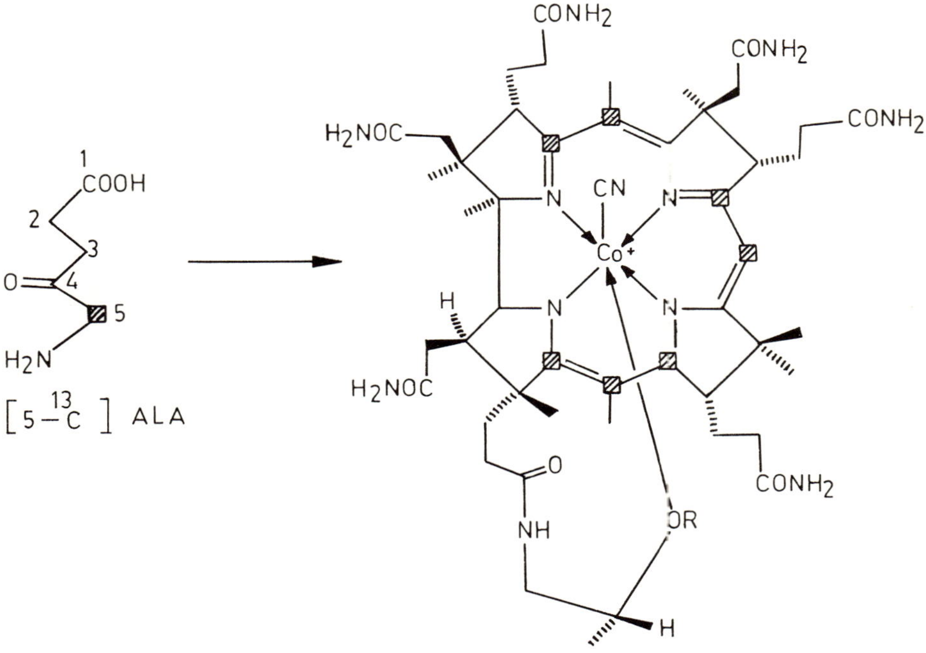

R = phosphate— ribose— DMBI

Since uroporphyrinogen III serves as a good precursor of Vitamin-B_{12} and since it has been demonstrated (A. R. Battersby, J. Moron, E. McDonald and J. Feeney, Chem. Comm., 920 (1972)) that it contains a C-5 ALA label in all four meso positions, one of these (that between rings A and D) must be lost in the subsequent mechanism as CH_2O or its equivalent (A. I. Scott, E. Lee and C. A. Townsend, Bioorg. Chemistry, 3, 229 (1974)). Essentially identical results using $5\text{-}^{13}\overline{C}$ ALA have been obtained by Shemin (C. E. Brown, J. J. Katz and D. Shemin, Proc. Nat. Acad. Sci., U.S.A., 69, 2585 (1972)).

11. The sequence of methylations with S-adenosylmethionine is based on rational mechanistic basis and not on results from experiments.

12. It is interesting to observe that alkylations take place from the α-side (excepting site 17), whilst prototropic shifts result in proton acceptance on the -β-side.

13. The pericyclic process envisaged involve 16 electrons leading to the observed conrotatory cyclisation. For a similar fascinating _in vitro_ cyclisation leading to the generation of the corrin nucleus see, Y. Yamada, D. Miljkovic, P. Wehrli, B. Golding, P. Loliger, R. Keese, K. Muller and A. Eschenmoser, Angew. Chem., Int. Ed. Engl., _8_, 343 (1969).

14. The intermediacy of cobyrinic acid was proved by the isolation of this compound from the incubated mixture of [5-^{14}C] ALA with a crude mixture of 'corrin synthetase' prepared from the wet cells of _P. shermanii_ (A. I. Scott, B. Yagen and E. Lee, J. Amer. Chem. Soc., _95_, 5761 (1973)).

15. The carboxyl groups at a, b, c, d, e and g can be amidated stepwise to yield the final hexamide cobyric acid. Cobyric acid has been isolated from _P. shermanii_ (K. Bernhauer, E. Becher, G. Gross and G. Wilharm, Biochem. Z., _332_, 562-72 (1960)). In an alternative pathway, the carboxyl group at position f can be amidated with 1-aminopropan-2-ol, to yield the product cobinic acid. The cobinic acid can then undergo stepwise amidation of the remaining carboxyl groups a, b, c, d, e and g to yield cobinamide. Once cobyrinic acid starts down one of the above two pathways, however, it is not committed to continue through to the products mentioned, but one can cross from one path to the other (B. F. Burnham in 'Metabolic Pathways', Vol. III, 3rd ed., D. M. Greenberg, Ed., Academic Press, New York, N.Y. 1969, Chapter 18.

16. It has been demonstrated that L-threonine is the precursor of the amino-2-propanol (A. J. Krasna, C. Rosenblum and D. B. Sprinson, J. Biol. Chem., _225_, 745 (1957)).

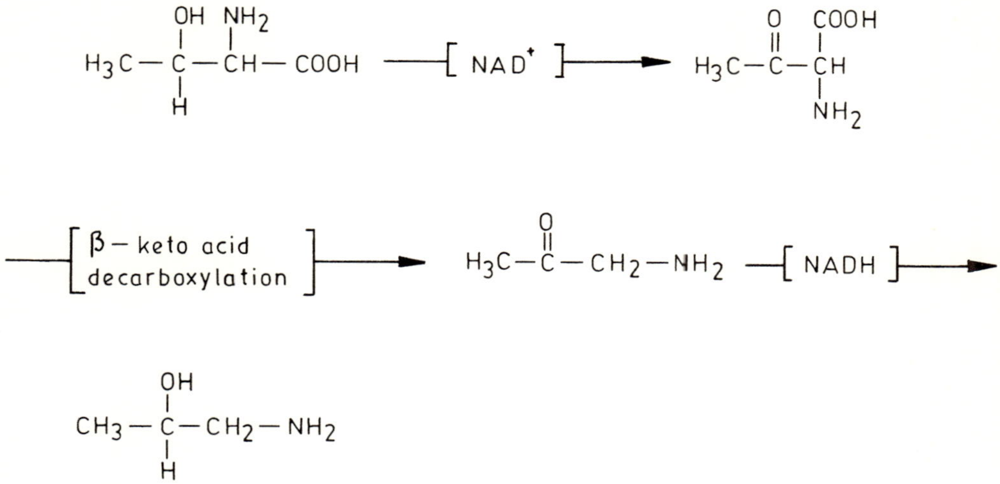

(J. M. Turner, Biochem. J., 99, 427 (1966).)

17. Cultures of P. shermanii, grown in the presence of cobalt but without precursor 5,6 dimethyl benzimidazole produce mainly cobinamide (K. Bernhauer, E. Becher, G. Gross and G. Wilharm, Biochem. Z., 332, 562 (1960); F. Wagner, Annu. Rev. Biochem. 35, 405 (1966)).

18. Cobinamide-phosphate has been identified in suspensions of vitamin B_{12}-synthesizing bacteria (R. Barchielli, G. Borretti, A. DiMarco, D. Julita, A. Migliacci, A. Minghetti and C. Spalla, Biochem. J., 74, 382 (1960)).

19. Cobinamide-guanosine-diphosphate has been characterised in P. shermanii (Ref. 18, B. Bartoskinski, B. Zagalsk and J. Pawelkiewicz, Biochim. Biophys. Acta, 136, 581 (1967)).

20. The biosynthesis of the dimethyl benzimidazole (DMBI) unit has been demonstrated to take place via an unusual and mechanistically fascinating sequence starting from guanine. The transformation of guanine to DMBI follows a pattern precisely similar to that established for the biosynthesis of riboflavin. The rationalisation proposed below for the transformation is also supported by in vitro experiments:

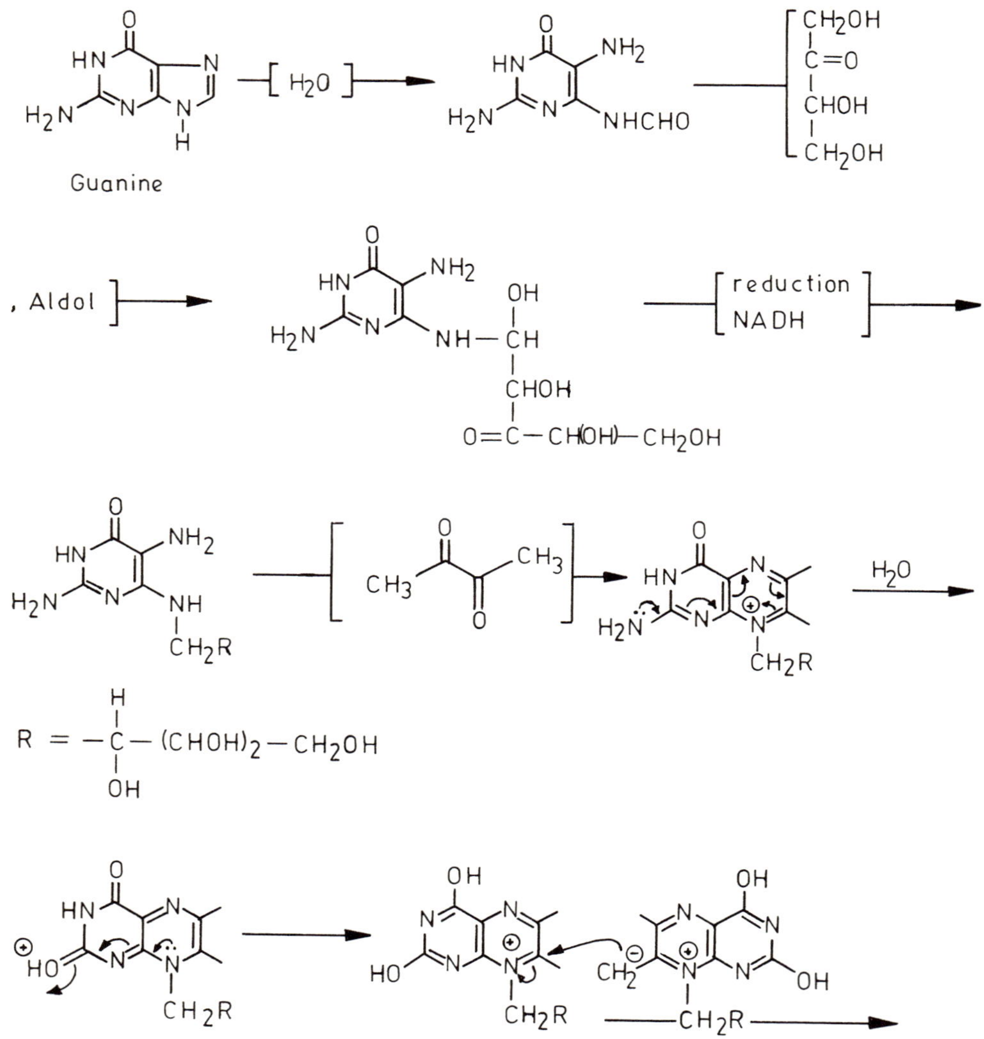
Guanine
H2O
, Aldol
reduction
NADH
H2O
R = —C—(CHOH)2—CH2OH

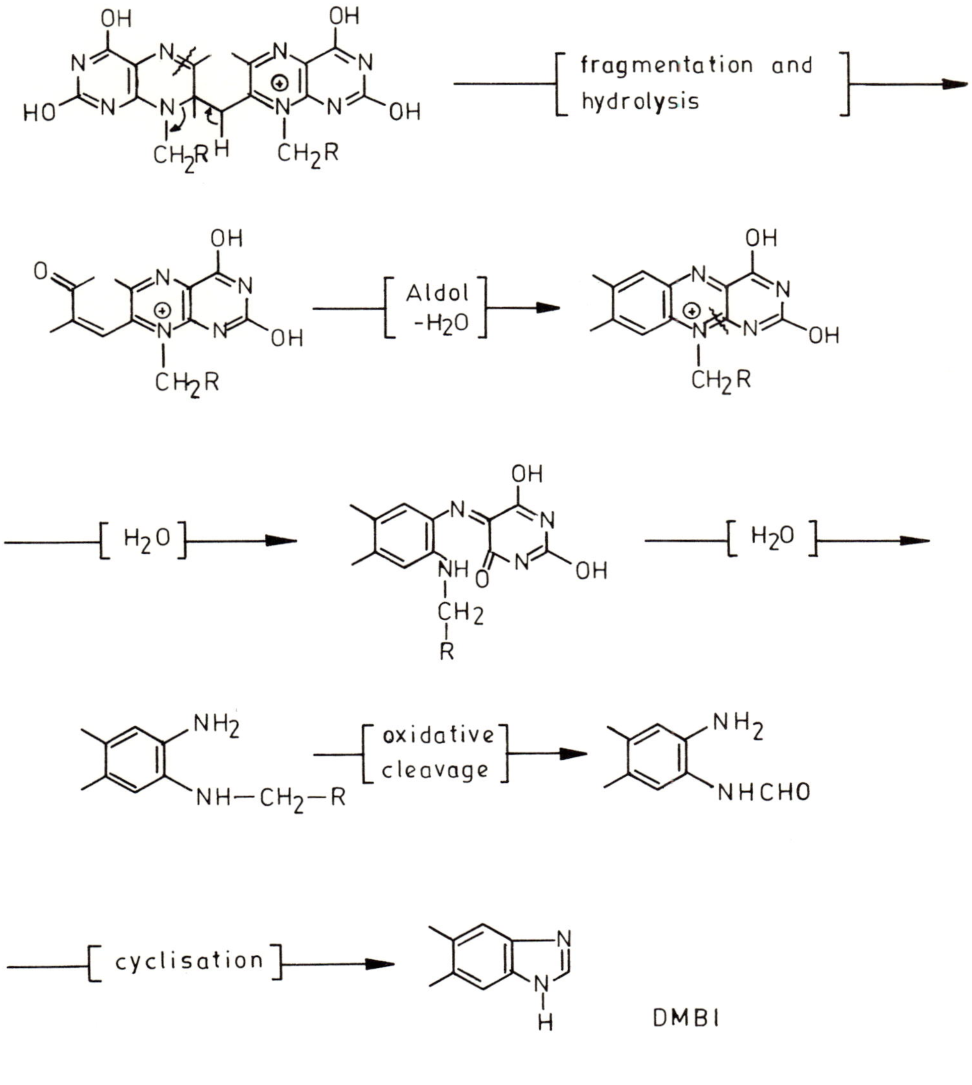
fragmentation and hydrolysis
Aldol
-H2O
H2O
H2O
oxidative cleavage
cyclisation
DMBI

(W. L. Alworth, S. H. Lu and M. F. Winkler, Biochemistry, 10, 1421 (1971); T. W. Goodwin in 'Metabolic Pathways', Vol. IV, 3rd ed., D. M. Greenberg, Ed., Academic Press, New York, N.Y., 1970, p. 353).

21. α-Ribazole 5' phosphate (1-α-D-ribofuranosyl-5,6-dimethyl-benzimidazole has been identified in *P. shermanii* (H. C. Friedmann and D. L. Harris, Biochem. Biophys. Res. Comm., 8, 164 (1962)).

22. Cobalamin-5'-phosphate has been found to occur in vitamin-B_{12} synthesising bacteria (H. C. Friedmann and D.L. Harris, J. Biol. Chem. 240, 406 (1965)). Enzymatic dephosphorylation gives the corresponding nucleoside.

23. It is interesting to observe that vitamin-B_{12} is not a natural product but arises with intense ease from the co-enzyme.

Glossary

*Anaerobic Glucose Metabolism

This is the most important pathway for the degradation of glucose. The net reaction involves transformation to pyruvic acid and its storage as lactic acid:

$$C_6H_{12}O_6 + 2ADP + 2\ H_3PO_4\ (Pi) \longrightarrow$$

$$2CH_3-\underset{\underset{OH}{|}}{CH}-COOH + 2\ ATP + 2\ H_2O$$

Lactic acid

only about 7% of glucose energy is transformed to energy rich ATP.

*ATP (Adenosine triphosphate)

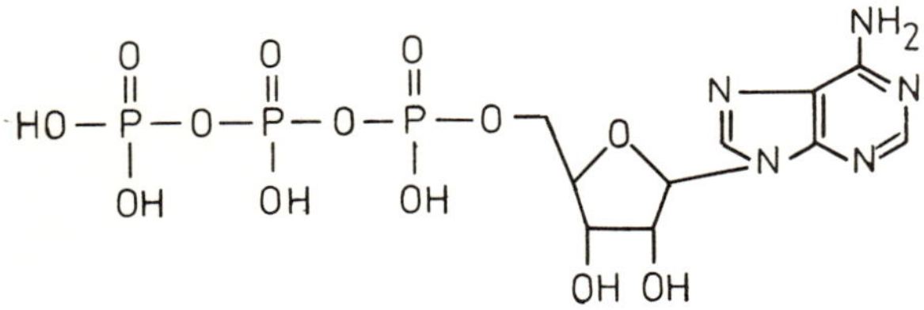

ATP can be correctly described as the reagent that drives life processes. The functional part is the energy rich triphosphate unit. The activation by ATP can be generalised

$$X + ATP \longrightarrow X^* + ADP/AMP$$

$$\text{OR} \quad X + ATP \longrightarrow X^* - P - P + AMP$$

In either case the substrate is activated for further reactions involving roughly a transfer of 7 kcal. bond energy. Thus a variety of sluggish functional groups such as carboxylic acids and their derivatives, alcohols and their derivatives are activated with ATP to form, inter alia, proteins, nucleic acids, polysaccharides and isoprenoids. The biodegradation of these molecules - taken as food - produces, in turn, the ATP. The ultimate source of ATP however is solar energy available by photosynthesis.

* Biotin

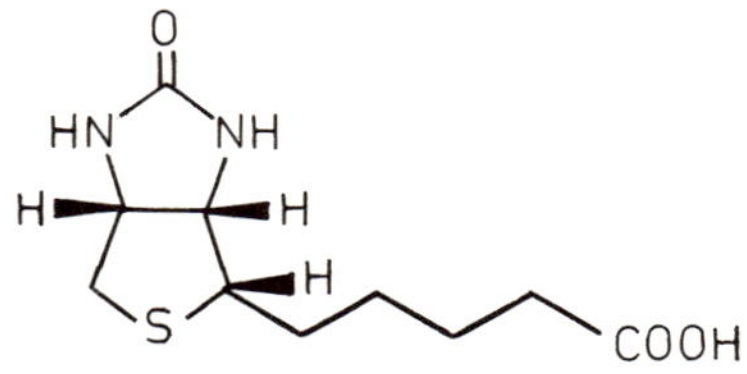

Biotin

Biotin mediates the entry of carbon dioxide into biological systems via adduct involving the basic ring nitrogen.

* <u>Coenzyme A (CoA-SH)</u>

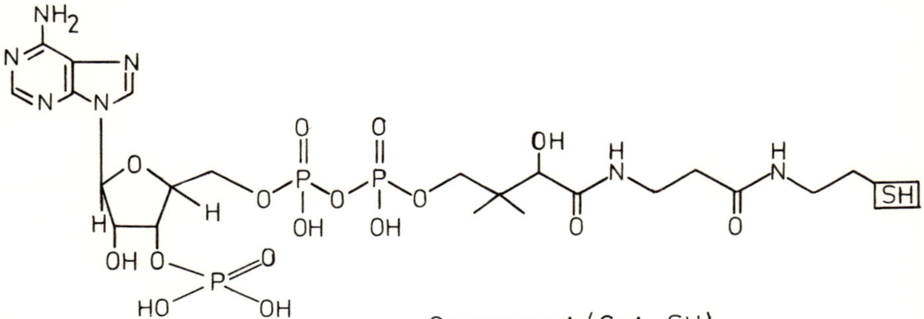

Coenzyme A (CoA-SH)

It is interesting to observe that Nature prefers to create molecular frameworks with thiol esters in contrast to the nearly exclusive <u>in vitro</u> use of alcohol esters. Such thiol esters are invariably those arising from coenzyme A (CoA-SH).

$$R-\overset{\overset{\displaystyle O}{\|}}{C}-X \; + \; CoA\text{-}SH \longrightarrow R-\overset{\overset{\displaystyle O}{\|}}{C}-S-CoA \; + \; XH$$

Thiol esters are 'activated' in the sense that they could be hydrolysed readily and are stronger acids compared to alcohol esters. These properties are related to the decreased contribution to the stabilisation of the carbonyl function by the sulphur lone pair compared to the oxygen analog:

$$R-\overset{\overset{\displaystyle O}{\|}}{C}-\ddot{\underset{..}{S}}-R' \longleftrightarrow R-\overset{\overset{\displaystyle O^{\ominus}}{|}}{C}=\overset{\oplus}{S}-R'$$

The higher acidity of thiol esters makes it considerably easier to bring about Claisen and related condensations under the mild biological conditions.

* <u>$Cl^{\oplus}$</u>

The formation of this unusual reagent - responsible for the genesis of a host of chlorine containing compounds - can be rationalised on the basis of oxidation of HCl:

$$RH + O_2 \longrightarrow R{-}O{-}OH$$

$$ROOH + HCl \longrightarrow ROH + H_2O + Cl^{\oplus}$$

The introduction of Br and I into biological substrates could also take place in a similar manner.

* <u>Flavin Adenine Dinucleotide (FAD)</u>

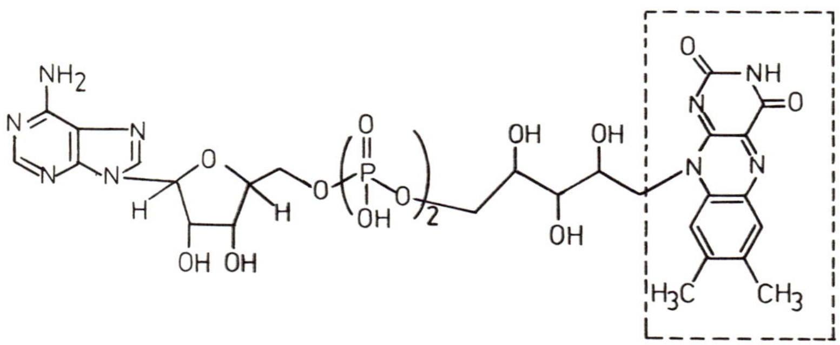

FAD Flavin

The heart of FAD is Flavin. This electrophilic substrate can accept two electrons by either polar or radical intermediates:

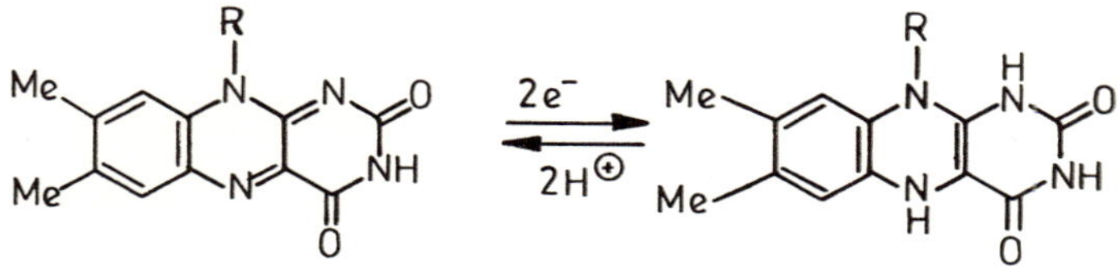

FAD FADH$_2$

FAD is actively involved in many oxidations such as

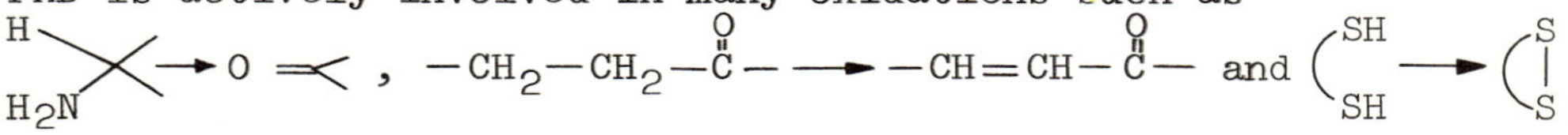

* <u>Ferredoxin</u>

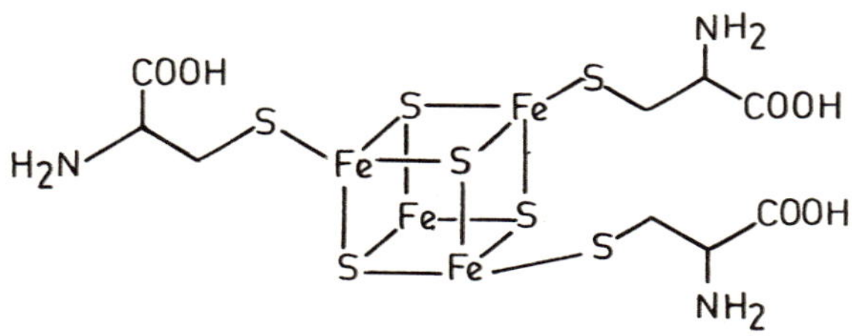

a model for Ferredoxin

Iron is our most abundant metal and Nature uses iron to create innumerable molecular frameworks. One fascinating aspect is the creation of diverse types of reagents from a central iron atom by a careful choice of ligands. Thus oxidising agents of varying shades like the cytochromes and the powerful reducing agent ferredoxin arises from the same Fe^{III}-Fe^{II} redox system. The sulphur ligands make the Fe^{II} state of ferredoxin very unstable thus making it a powerful reducing agent in various biological processes such as photosynthesis and nitrogen fixation.

* FH_4 (Folic acid, tetra hydro)

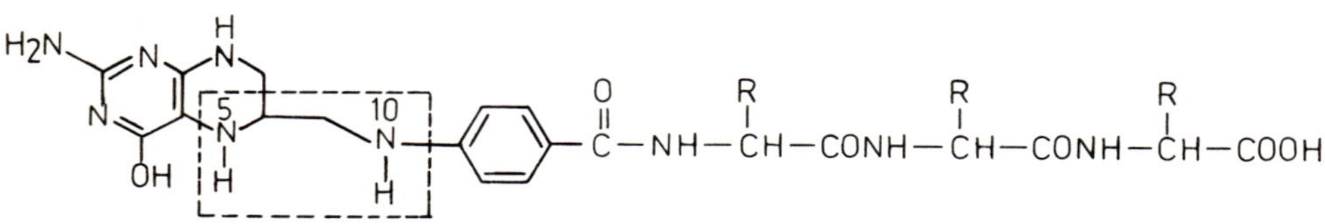

FH_4 R = $-CH_2-CH_2-COOH$

The heart of FH_4 is the 5,10 dinitrogen jaw which can sequester and release equivalents of formaldehyde, formic acid and methanol:

a: "formaldehyde" transfer:

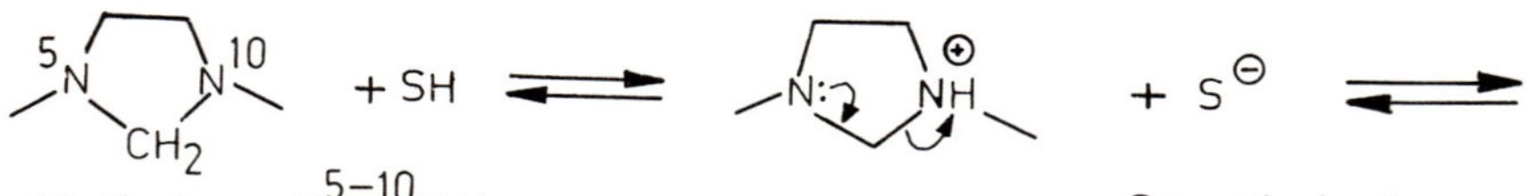

Methylene $N^{5-10}FH_4$ S = substrate

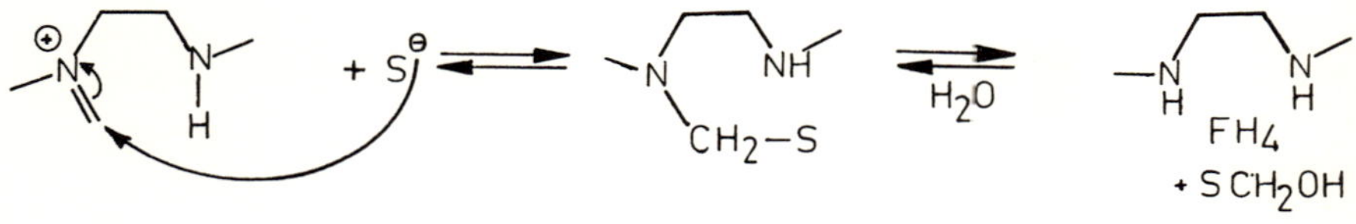

b: "formic acid" transfer

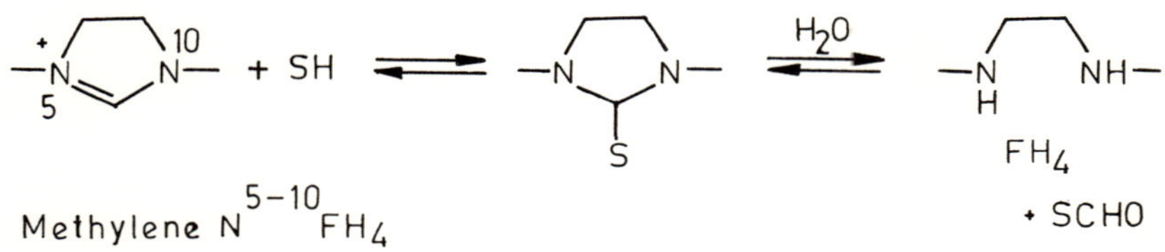

S = substrate

c: "methyl" transfer

S = substrate

Methyl N^5 FH_4

* Lipoic acid

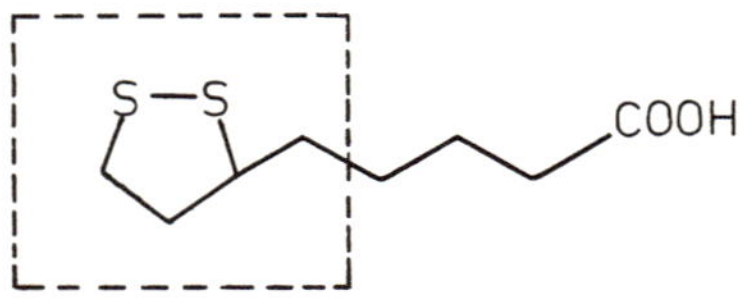

Lipoic acid

Lipoic acid effectively brings about the equivalent of

$R-\overset{\ominus}{C}=O \rightarrow R-\overset{\oplus}{C}=O$ change!

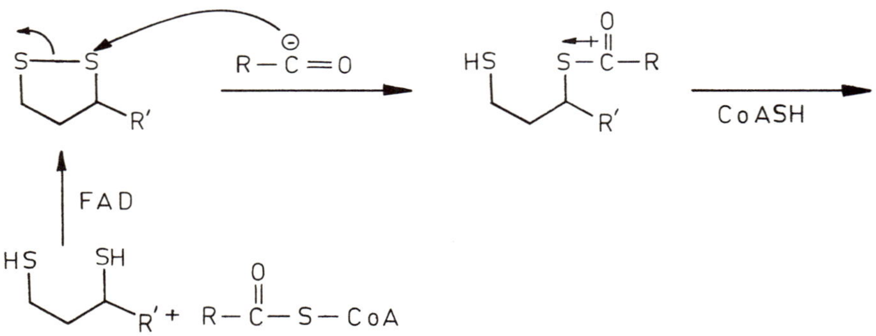

* Malonyl-CoA

$$HOOC-CH_2-\overset{O}{\overset{\|}{C}}-S-CoA$$

This synthon is responsible for the construction of polyketide units, related to the biosynthesis of fatty acids, aromatic compounds and several others:

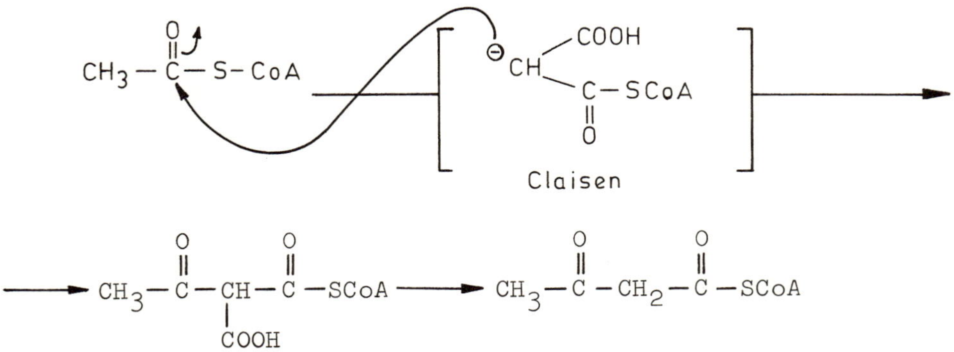

The net reaction really involves the equivalent of a Claisen condensation with acetyl-CoA alone; obviously the increased acidity of malonyl CoA makes it more efficient than acetyl-CoA in condensation process.

<u>NADH</u> (<u>Nicotinamide adenine dinucleotide, reduced</u>)

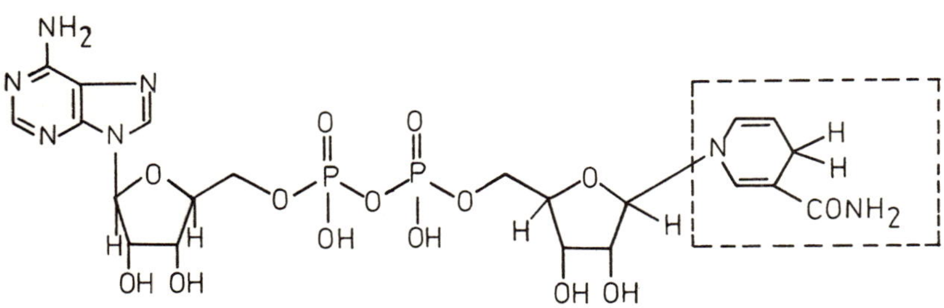

NADH

The NAD $^{\oplus}$—NADH (DPN $^{\oplus}$—DPNH) and the closely related NADP $^{\oplus}$—NADPH (TPN $^{\oplus}$—TPNH) are associated with practically every type of oxidation-reduction process associated with life. They involve the net transfer of two electrons:

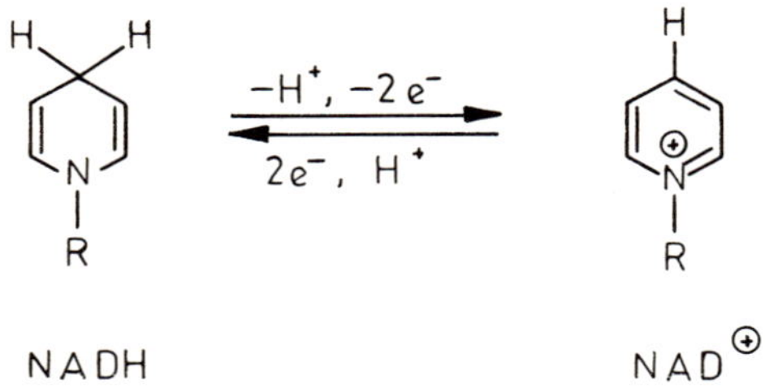

The NADH→NAD$^+$ change is a favourable one and this aspect has been taken advantage by Nature in utilising this system not only for redox reactions but also as a means to store and release energy.

* <u>OH</u> $^{\oplus}$

Electron rich substrates accept electrophilic oxygen, symbolised as OH$^+$, arising from induced heterolytic cleavage of the peroxide bond.

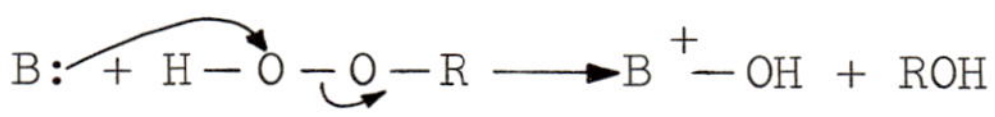

* <u>Pentose shunt</u>

Pentose shunt, an alternate pathway for glucose metabolism, is characterised by the formation of NADH and a fascinating array of molecular reorganisations.

* Pyridoxal - Pyridoxal phosphate

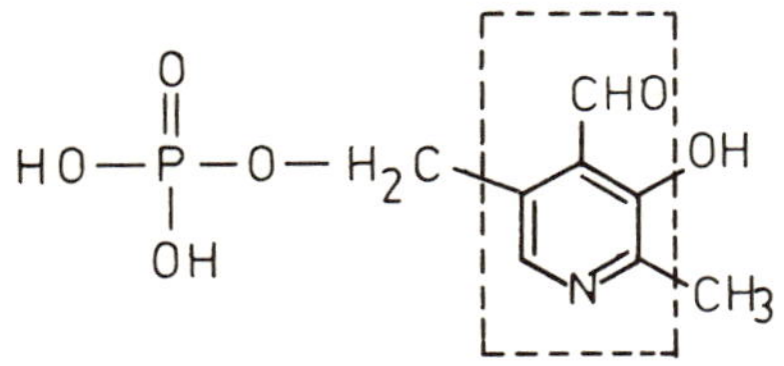

pyridoxal phosphate

Pyridoxal is the most striking of the known co-enzymes in terms of multiplicity of different biological reactions that are dependent on its presence. Thus α-amino acids could be racemised and deprived of either the carboxyl or the amino function or both by pyridoxal. Additionally, pyridoxal is involved in the α, β and β, γ elimination of appropriate α-amino acids. Such diverse transformations, mediated by the appropriate enzymes, are initiated by the expected Schiff base formation and

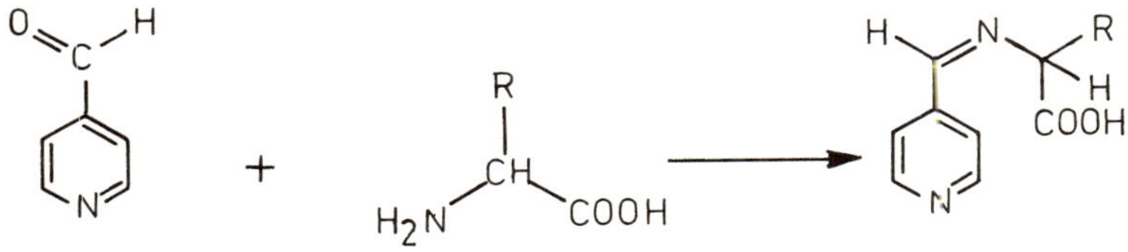

proceed via a fascinating sequence which highlights the beautiful molecular design that is inherent in this reagent!

a. Transamination and racemisation

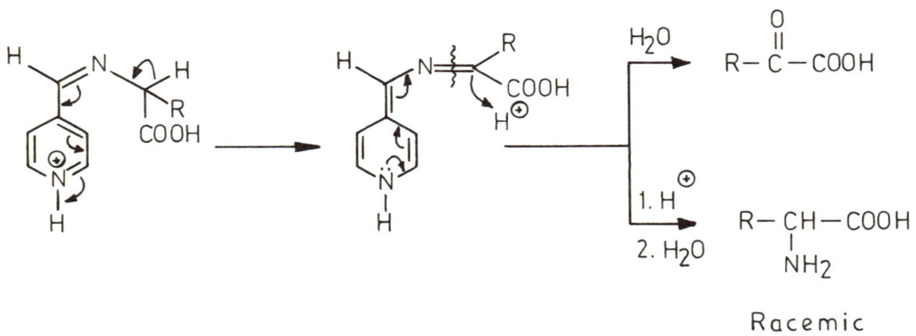

b. Decarboxylation-deamination and decarboxylation:

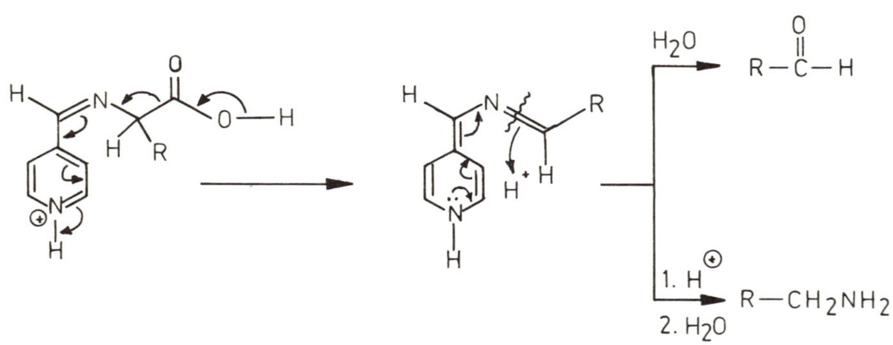

* Thiamine Pyrophosphate

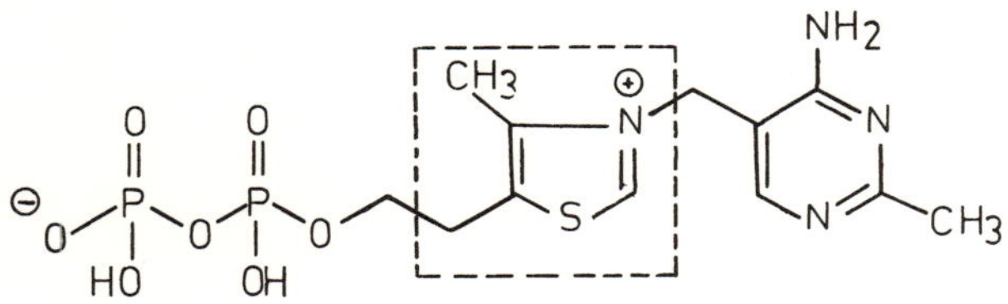

Thiamine pyrophosphate (TPP)

Biological processes are mediated by a small number of highly versatile reagents in whose choice is often reflected the profound experience of Nature. Thiamine pyrcphosphate above others highlights this facet. The N-alkylated thiazole system in TPP is a relatively strong acid with the conjugate base possessing qualities as an electrophile and a nucleophile!

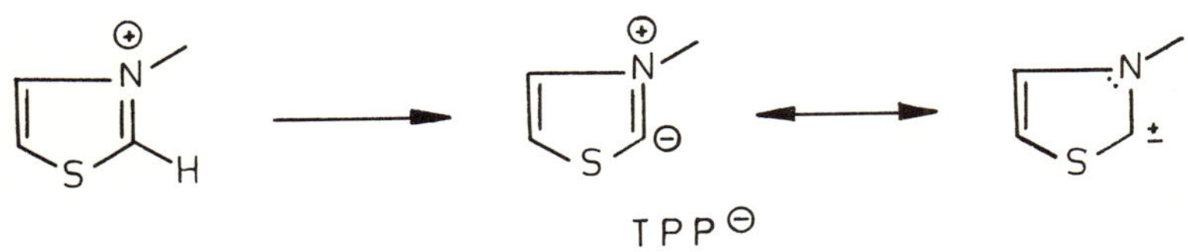

Transformations brought about by TPP $^{\ominus}$–α-keto acid decarboxylation, 'benzoin type' condensations - have practically no parallel in organic chemistry and involve the equivalent of $R-\underset{\ominus}{C}=O$ as intermediates:

a. α-Keto acid decarboxylation

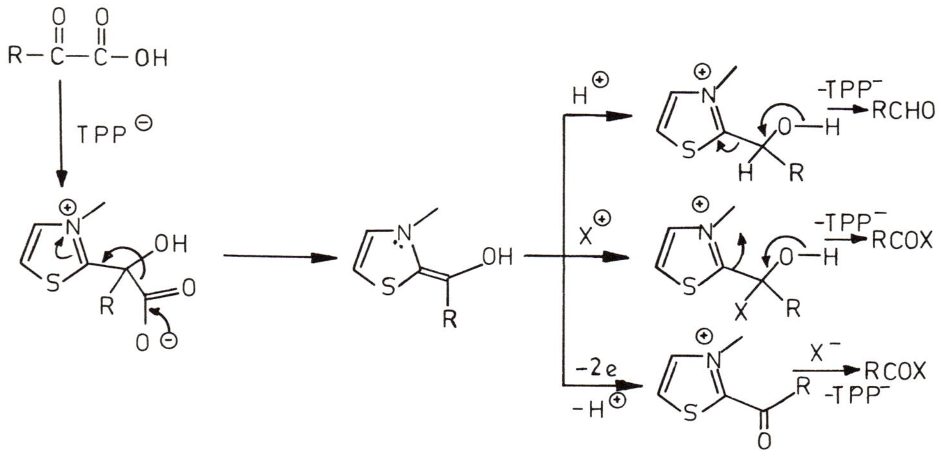

b. 'benzoin type' condensation

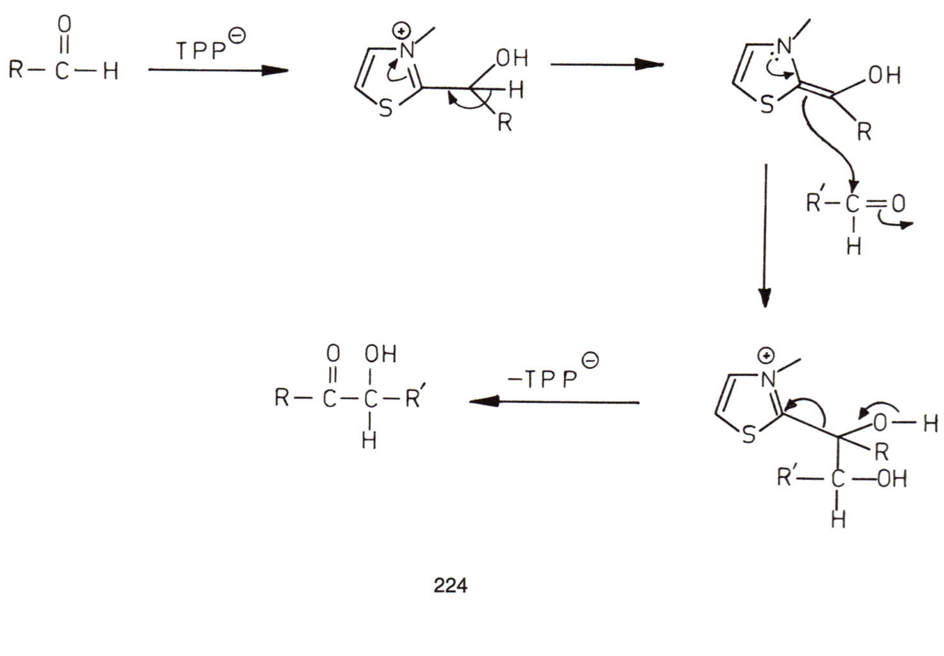

*Vitamin-B_{12}

Vitamin-B_{12} perhaps is the most enigmatic reagent of Nature. The great majority of the reactions mediated by Vit-B_{12} has no parallel in *in vitro* systems. The extraordinary properties of this coenzyme could be related to the Co^{III}—Co^{I} system that is present.

a. process initiated by proton loss from carbon acid:

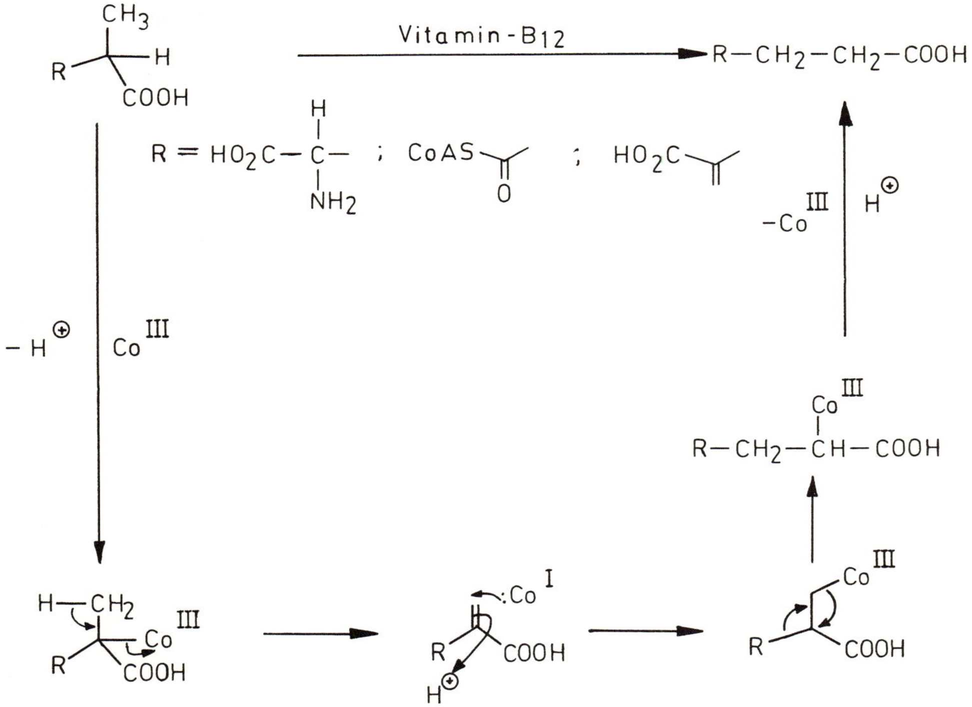

b. process initiated by hydride abstraction

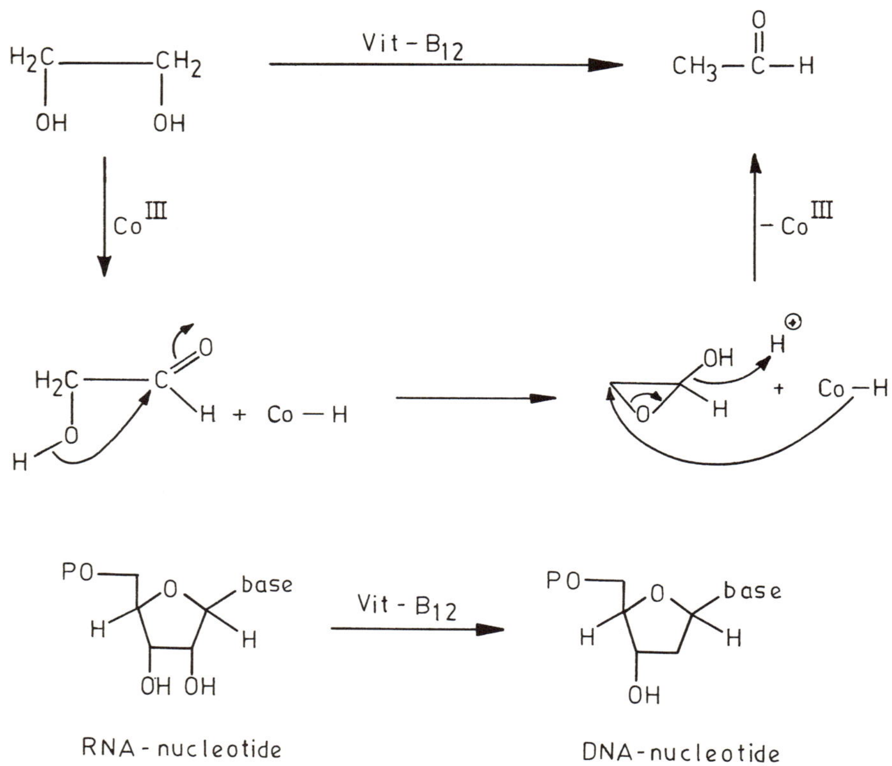

c. Vitamin-B_{12} as a "CH_3 (±)" Donor!

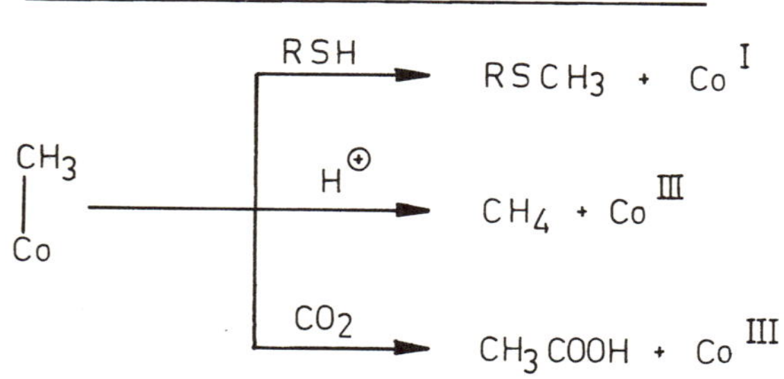

Author Index

Subject Index

α-Amino Acid Index

Reagent Index

Reaction-Type Index

A 6 B 7 C 8 D 9 E 0 F 1 G 2 H 3 I 4 J 5